Distributed Cooperative Control
and Communication for Multi-agent Systems

Dong Yue · Huaipin Zhang · Shengxuan Weng

Distributed Cooperative Control and Communication for Multi-agent Systems

Springer

Dong Yue
Nanjing University of Posts
and Telecommunications
Nanjing, Jiangsu, China

Huaipin Zhang
Nanjing University of Posts
and Telecommunications
Nanjing, Jiangsu, China

Shengxuan Weng
Nanjing University of Posts
and Telecommunications
Nanjing, Jiangsu, China

ISBN 978-981-33-6717-3 ISBN 978-981-33-6718-0 (eBook)
https://doi.org/10.1007/978-981-33-6718-0

This Springer imprint is published by the registered company Springer Nature Singapore Pte Ltd.
The registered company address is: 152 Beach Road, #21-01/04 Gateway East, Singapore 189721, Singapore

Preface

With the popularization of intelligent devices and the rapid development of communication technology, modern control systems exhibit two prominent features, namely structural interactivity and functional complexity. The former roots in the couplings of systems while the latter represents the complex task to be realized. In this case, multi-agent systems (MASs) are introduced to address these two issues. In general, MASs consist of a group of connected autonomous agents integrating distributed sensing, communication, computing, and control in a networked environment, in which these agents can cooperate with each other and share information with their neighbors to accomplish a complex task together. Compared with an individual agent, MASs have several advantages, such as reliability, adaptivity, robustness, scalability, low cost of operation and maintenance, etc. Recently, MASs and their concomitant cooperation problems have received considerable attention in many fields, such as biology, physics, artificial intelligence, and engineering.

To achieve a desired global goal, two interesting and crucial issues will be considered in multi-agent cooperation. One is system-level question, which involves how to coordinate and control such systems or agents; The other is network-level question related to how information among agents could be transmitted, especially in an efficient way since communication resources are often limited. The former question comes down to cooperative control protocol design while the latter to cooperative communication scheme design. Thus, these two aspects, which are important handgrips to deal with cooperation problems of MASs, will be investigated in this book. In general, cooperative control aims to drive a group of agents to reach an agreement on a common quantity of interest. A key issue of cooperative control is how to design an information flow algorithm which describes the rules of information interaction between adjacent agents. Depending on the differences of information interaction and processing unit of data fusion, cooperative control can be classified into three configuration modes: centralized, decentralized, and distributed ones. In large-scale MASs, centralized framework increases communication cost and computing burden while decentralized scheme lacks the information interaction among agents to achieve cooperation. Thus, it is preferable to employ distributed cooperative control which depends only on local interactions among the neighbor agents. This monograph mainly focuses on the various classes of distributed MASs cooperation problems.

During the design and implementation of distributed cooperative control, constrained communication resources of the shared network play a vital role. In general, time-triggered communication scheme is an underlying assumption in a conventional cooperative control setting. Yet time-triggered strategy, in which agent accesses the continuous information of its neighbors periodically, may lead to heavy consumption of communication and computation, especially when the system tends nearly to be stable or no disturbances are imposed on the system. Thus, it is urgent to design a suitable communication scheme which can reduce the consumption of communication resources while ensuring the desired performance of MASs. Event-triggered communication mechanism is introduced into the cooperation problems of MASs. In an event-triggered scheme, the time instants when data transmissions and control updates of the agents are judged by the predesigned trigger conditions depending on the measurements of itself and its neighborhoods. These two cooperative communication scheme will be another research topic of this book.

The books are structured as follows: (1) Chap. 1 presents the introduction of the developments of distributed MASs cooperation and analyzes cooperative control method and communication framework of MASs; (2) Chaps. 2 and 3 focus on distributed cooperation problem of linear homogeneous and heterogeneous MASs; (3) The consensus problems of nonlinear MASs with communication delays and state saturations are investigated in Chaps. 4 and 5; (4) The cooperation framework is applied to multiple rigid body systems in Chaps. 6 and 7; (5) Data-driven method is employed to address the cooperative control problem while minimizing the system performance in Chaps. 8 and 9.

We are indebted to the collaborations with Dr. Xiuxia Yin, Yang Yang, and Hongjun Chu on the work of linear and nonliear MASs cooperation. The work was supported by the Natural Science Foundation of Jiangsu Province of China under grant BK20202011, the National Key R&D Program of China under grant 2018YFA0702200, the National Natural Science Foundation of China under grant 61533010, 61903199 and 62073170, and the Natural Science Foundation of Colleges and Universities of Jiangsu Province under grant 19KJB120004.

Nanjing, China
September 2020

Dong Yue
Huaipin Zhang
Shengxuan Weng

Contents

1 Overview of Multiagent Systems Cooperation 1
1.1 Introduction 1
1.2 Cooperative Control of Multiagent Systems 1
1.2.1 Model-Based Cooperative Control Method 2
1.2.2 Data-Driven Cooperative Control Method 3
1.3 Cooperative Communication of Multiagent Systems 4
1.4 Contribution of This Book 5
1.5 Book Organization 7
References 9

Part I Distributed Cooperative Control of Linear Multi-agent Systems

2 Distributed Event-Triggered Consensus Control of Homogeneous Multiagent Systems 17
2.1 Problem Formulation 17
2.2 Distributed Consensus Control Design and Analysis 19
2.2.1 Distributed Adaptive Event-Triggered Strategy 19
2.2.2 Distributed Consensus Control Design 20
2.2.3 Modeling of The Closed-Loop Systems 24
2.2.4 Consensus Analysis 27
2.3 Uniform Delay Case 33
2.4 Simulation Example 36
2.5 Conclusion 40
References 40

3 Distributed Event-Triggered Tracking Control for Heterogeneous Multiagent Systems 43
3.1 Problem Formulation 43
3.2 Event-Triggered Tracking Control Design 44
3.3 Consensus Analysis 48
3.4 Simulation Example 61

3.5 Conclusion 63
References 64

Part II Distributed Cooperative Control of Nonlinear Multi-agent Systems

4 Distributed Formation Tracking Conrol of Nonlinear Mutiagent Systems 67
4.1 Problem Formulation 67
4.2 Formation Tracking Control Strategy Design 69
4.3 Simulation Examples 75
4.3.1 A Team of Mobile Robots 75
4.3.2 Numerical Example 79
4.4 Conclusion 81
References 82

5 Distributed Consensus Control of Nonlinear Multiagent Systems 83
5.1 Problem Formulation 83
5.2 Continuous Consensus Protocol Design 85
5.3 Intermittent Consensus Protocol Design 91
5.4 Adaptive Consensus Protocol Design 93
5.5 Extension to Output Feedback Case 94
5.6 Applications in Constrained Preserving Connectivity 96
5.7 Simulation Examples 98
5.7.1 Application to Aircraft Team 99
5.7.2 Application to Connectivity Preservation 100
5.8 Conclusion 106
References 106

Part III Distributed Cooperative Attitude Control of Multiple Rigid Body Systems

6 Distributed Event-Triggered Attitude Cooperative Control of Multiple Rigid Body Systems 109
6.1 Problem Formulation 109
6.2 Inter-Group Anti-consensus Algorithms 113
6.3 Intra-Group Event-Triggered Synchronization Algorithms 115
6.4 Simulation Example 121
6.5 Conclusion 125
References 126

7 Distributed Robust Attitude Containment Control for Multiple Rigid Body Sysems 127
7.1 Problem Formulation 127
7.2 Attitude Containment Control Law Design 129
7.2.1 The Sliding-mode Surface Design 129

7.2.2 Sliding-mode Attitude Containment Control Design 134
7.2.3 Adaptive Sliding-mode Attitude Containment Control Design 139
7.3 Simulation Examples 143
7.3.1 Sliding-mode Attitude Containment Control 144
7.3.2 Adaptive Sliding-mode Attitude Containment Control 146
7.4 Conclusion 148
References 148

Part IV Distributed Data-Driven Optimal Consensus Control for Multi-agent Systems

8 Data-Driven Optimal Consensus Control for Linear Multiagent Systems 151
8.1 Problem Formulation 152
8.2 Problem Transformation 153
8.3 Optimal Consensus Control Design 160
8.3.1 Optimal Consensus Control for Delay-free Multiagent Systems 160
8.3.2 Nash Equilibrium 162
8.4 Policy Iteration Algorithm 163
8.5 Data-driven Approximation Solutions for The Coupled HJB Equations by Critic-Actor Neural Networks 165
8.5.1 Critic Network 165
8.5.2 Actor Network 166
8.6 Simulation Example 167
8.7 Conclusion 169
References 171

9 Data-Driven Event-Triggered Optimal Consensus Control for Nonlinear Multiagent Systems 173
9.1 Problem Formulation 173
9.2 Event-Triggered Optimal Consensus Control Design 175
9.3 Online Solution of Optimal Consensus Controllers Using Neural Networks 176
9.3.1 ETADP Approach 177
9.3.2 Smart Critic-Action Neural Networks Design 178
9.3.3 Stability Analysis 181
9.4 Simulation Example 187
9.5 Conclusion 190
References 191

Appendix: Algebraic Graph Theorys 193

Index 195

Acronyms

$\forall$	For all
$\in$	Belong to
$\sum$	Summation
$\otimes$	Kronecker products
$\cup$	Union operator
$\circ$	Compound operator
sup	Supremum
inf	Infimum
R	Set of real numbers
R^n	n dimensional real Euclidean space
$R^{n\times m}$	$n \times m$ dimensional real matrix space
A	Real matrix A
A^T	Transpose of matrix A
A^{-1}	Inverse of matrix A
$\lambda_i(A)$	The ith eigenvalue of matrix A
$\lambda_{max}(A)$	The maximum eigenvalue of matrix A
$\lambda_{min}(A)$	The minimum eigenvalue of matrix A
$A > 0$	Symmetric positive definite A
$A \geq 0$	Symmetric positive semi-definite A
$\mid x \mid$	Absolute value of x
$\|x\|_p$	p-norm of vector x
$\|A\|_p$	Induced p-norm of matrix A
$0_{n\times m}$	$n \times m$ zero matrix
I_n	$n \times n$ identity matrix

Chapter 1
Overview of Multiagent Systems Cooperation

1.1 Introduction

During the rapid and sustained development of industrial and civilian demand, modern control systems are becoming more and more open, complex, heterogeneous and highly distributed. Under this circumstance, it is expected to build new engineering styles to achieve many complex coordination tasks, which is difficult to be implemented based on individual agent. Multi-agent system (MAS), which imposes a group of connected autonomous agents operating through distributed sensing, communication, computing, and control in a networked environment to achieve a certain system goal, is regarded as a promising and effective system architectural to cope with these coordination problems. In the network of MAS, each agent is described by a node and the connections among agents are represented by edges. To cooperate with other agents, each agent needs to share information with its neighbor agents through the edges so as to reach agree on a common quantity of interest. In comparison with an individual agent, MASs have some prominent advantages, including strong robustness and reliability, high scalability and efficiency, manipulability, large-scale spatial distribution. MASs have attracted considerable research attention from various fields, including biology, computer science, robotics and engineering during the last decade, this is mainly due to the widespread application in scientific computing [1, 2], distributed artificial intelligence [3, 4], space exploration [5, 6], mobile robotics [7, 8], power systems [9, 10], search and rescue [11, 12].

1.2 Cooperative Control of Multiagent Systems

In the control design of MASs, the main objective is to make all agents operate in a cooperative fashion through a control protocol. Here, cooperative refers to a close relationship among agents, where all agents can achieve a certain common

D. Yue et al., *Distributed Cooperative Control and Communication for Multi-agent Systems*, https://doi.org/10.1007/978-981-33-6718-0_1

agreement. Depending on the difference of coordinated tasks and interaction communication of agents, cooperative control is roughly classified into three categories: consensus control, formation control, containment control. Numerous results concerning cooperative control have appeared [13–17] recently. A key issue on cooperative control is how to design an appropriate control protocol by making use of global/local interaction information among the neighboring agents to achieve a collective behavior. In recent, an amount of research efforts have been made in analyzing the control method for multi-agent cooperative control, in which they are classified into two types: model-based control approaches and data-driven control techniques. More details are descried as following.

1.2.1 Model-Based Cooperative Control Method

The traditional approaches of designing MASs usually involve developing a mathematical model describing the dynamics of the agents, and then applying analytical and synthesis techniques to the development model to derive a suitable stability control protocol. In general, such a mathematical model of MAS consists of state-space equations, most of which are obtained using some identification and estimation methods. This approach is often termed model-based control which has been fully developed to be a relatively mature theory. Numerous research on model-based approaches for cooperative control of MASs are found in [13, 14, 18, 19].

For linear MASs, Olfati-Saber and Murray in [20, 21] introduced the theoretical framework of model-based consensus control problems. To judge the existence of consensus under a model-based control protocol, the consensusability problem was presented in [22–24]. The authors of [25–28] designed the model-based cooperative control protocols for both the first-order continuous-time and discrete-time MASs. In [29, 30], Li et al. employed H_∞ performance allocation method to study cooperative output regulation problems of heterogeneous MASs. In [31–33], observer-based leader-following consensus problems of MASs was investigated, where various distributed model-based consensus control protocols are proposed. Dong in [34] considered consensus issue of high-order nonlinear MASs with uncertainty, distributed dynamic consensus controller was designed with the aid of an inner-loop and outer-loop approach.

Extensive efforts of model-based formation control design of MASs were found in [35–39]. In [40], a comprehensive review on model-based formation control strategies of multiple unmanned ground vehicles was presented. To guarantee the transient performance of the closed-loop systems, Mehdifar et al. [41] proposed decentralized robust distance-based formation control algorithm of MASs affected by unknown external disturbances. In constrained environments, model-based optimal dynamic formation issue of leader-follower MASs was addressed in [42]. [43, 44] focused on output regulation issues of heterogeneous MASs.

The containment control problems of continuous-time and discrete-time MASs were discussed in [25, 45, 46], in which distributed control laws based on the relative states and outputs of neighborhood are constructed. In [25, 47], adaptive and asynchronous output containment control of heterogeneous MASs were considered. The analysis on distributed finite-time attitude containment control for multiple rigid bodies was presented in [48, 49], which can drive the attitudes of the followers into the convex hull formed by the one of leaders. Qi et al. [50] proposed a distributed event-triggered fuzzy bipartite containment controller of nonlinear MASs with input quantization over a signed digraph, in which a fuzzy observer was designed to estimate unmeasured states.

1.2.2 Data-Driven Cooperative Control Method

In the past few decades, the ever-increasing complexity of modern industrial systems has made it more and more difficult to build the accurate physical models using the first principles. Even though the physical models of the plants can be obtained, it may be difficult to use model-based control theory for controller design and performance evaluation since the models are too complex. With the rapid development of information technology, modern industrial processes can easily collect a large number of measured process data which contain stored historical data and online real-time data. In the case of failure of model-based control theory, how to use these measurable data for controller design, state prediction and performance evaluation is an important and urgent problem. In order to solve this issue, data-driven control, which can directly utilize measurable input/output data to design the controller, was introduced in [51–54]. In what follows, two typical data-driven methods will be utilized to study the cooperation control problems.

(i) Iterative learning control

Iterative Learning Control (ILC) is an intelligent learning control methodology which can improve the transient performance of the system in the current iteration by using the past control experience. It can effectively deal with repetitive control process in a finite duration and requires only process measurements instead of accurate system model. Recently, ILC has been extensively applied to study the cooperative control problems [55–63]. In [55], Ahn and Chen introduced ILC to consider the formation control problem of MASs with continuous nonlinear dynamics. Then formation-based ILC were extended to investigate MASs with switching topologies and actuator faults in [56, 57, 60]. Bu et al. [62] employed the terminal ILC method to discuss the finite-time consensus problem of nonlinear MASs with unknown dynamics and output saturation, where data-driven consensus protocols are developed. In [58] and [59], formation tracking control of second-order and high-order MASs were studied based on ILC approach. Hua et al. [63] introduced an event-triggered mechanism into ILC framework to solve the containment control problem of model-free MASs.

(ii) Adaptive dynamic programming

Adaptive dynamic programming (ADP) integrating reinforcement learning and adaptive control has provided an effective way to learn the HJB equations forward-in-time. Recently, ADP technique has been extensively used in MASs [64–71] which showed that it is a promising method to deal with optimal consensus control problems. In [67], an online algorithm was proposed to address optimal coordination control problem for nonlinear MASs through fuzzy ADP. References [64, 65] adopted ADP method to obtain adaptive learning solutions for multi-agent graphical games based on policy/value iteration algorithms. In [66, 72, 73], distributed adaptive output feedback controllers based on ADP have been developed to treat with output regulation of linear heterogeneous MASs. In order to address optimal consensus problem for MASs with external disturbances, [74] proposed a distributed adaptive optimal tracking controller by ADP technique.

1.3 Cooperative Communication of Multiagent Systems

In the analysis and synthesis of distributed cooperative control protocol, communication mechanism and controller actuation scheme are the prominent roles. In general, there are two scheduling and controlling scheme: time-triggered strategy [13, 75–78] and event-triggered strategy [14, 79–81]. In the time-triggered control, the controller updates of agents and information interaction between agents are carried out periodically. This kind of triggering mechanism may generate many unnecessary controller actuation and data transmission, which will increase the computational burden and communication pressure of the system. To mitigate the load caused by high frequency of transmitting and control executions of time-triggered method, event-triggered control mechanism is introduced in [80].

Recently, event-triggered control (ETC), which advocates the implementation of data transmission and control updates only when the predefined rules are violated, has attracted considerable attention [14, 17, 81–83]. In the ETC, the fundamental rationale is that how to judge the triggering instants depending on a predesigned triggering mechanism which is closely related to the agents states [80]. When the designed triggered condition is violated, the current states of the system should be transmitted and the controllers are updated. In this case, data transmission and control updates are performed only when needed, thus the ETC approach can effectively reduce the unnecessary utilization of limited communication and computation resources compared to time-triggered control [81].

Since Dimarogonas et al. [83] introduced the ETC method to MASs, the research on event-triggered cooperation control has achieved fruitful results. In [83], the centralized event-triggered controller was designed for first-order MASs in which the state information of all agents was required to a centralized decision maker and then the controllers were updated simultaneously. In addition, a distributed fashion was also introduced into ETC depended only on the information of itself and the neighbor agents which can further reduce data transmission and control updates. Some related results on single-integrator MASs [84–86] and double-integrator MASs [87, 88] can be found.

1.4 Contribution of This Book

This book investigates the distributed cooperation problems of MASs which mainly considers two factors that affect cooperation: distributed controller design and suitable communication strategy design. Thus distributed cooperation control design analysis of linear, nonlinear MASs and multiple rigid bodies are the first contribution of this book. Secondly, the proposed two kinds of communication strategies are another contribution. The more detailed contributions of this book are summarized as follows:

- By combining the distributed event-triggered communication scheme together with the networked networked predictive control scheme, distributed model-based event-triggered predictive control (D-MB-ETPC) method is proposed. The current states of agents are predicted efficiently based on available historical local event-triggered information and so the predictive controllers can be constructed. The proposed D-MB-ETPC has the advantages of reducing the utilization of the limited network resources significantly, compensating for the networked communication delays actively, and guaranteeing the desired consensus control performance.
- A novel distributed event-triggered communication scheme of discrete-time heterogeneous MASs with random communication delays is proposed. Sufficient conditions is developed to guarantee the consensus in mean square sense. Compared with the existing method, the designed event-triggered functions have several parameters that can be adjusted to reduce the communication burden and preserve the consensus performance of the MASs investigated.
- A time-varying formation tracking strategy is presented for uncertain nonaffine nonlinear MASs with communication delays. Based on active disturbance rejection control method, the nonaffine nonlinear MASs are transformed into affine nonlinear ones with uncertainties. Then the proposed strategy compensates uncertainties, in real time, by estimations of the extended states and it provides an effective method for the MASs information tracking control. The problem of explosion of complexity is overcome in formation tracking control of higher-order MASs, and the tracking differentiators estimate the derivatives of the desired signal and the virtual control variables. Unlike the traditional backstepping-based formation control schemes, the TDs are introduced to eliminate the requirement of the operations of higher-order derivatives of virtual control variables.
- Three types of protocols, namely continuous,intermittent, and adaptive state feedback protocols are proposed for achieving consensus of nonlinear MASs subject to the relative state saturations. Then consensus of general linear MASs with Lipschitz nonlinearity is extended to the case with relative state saturations and

stability of a single system under state saturation to consensus of MASs under relative states saturation. In contrast with the existing consensus problems which the relative states are constrained by hyperbolic tangent function, we will propose three types of state feedback protocols and the extension to output feedback case, with relative states being constrained to a hypercube. The resulting challenge of this extension is how to translate consensus with the relative state saturations into stability of edge dynamics systems operating on a constrained set, and how to construct saturation functions and controllers for stabilizing the edge systems subjected to a constrained set.

- Distributed event-triggered cooperative attitude control protocol of multiple groups of rigid bodies evolved on nonlinear manifold SO(3) is developed. Different from the conventional attitude synchronization, the multiple attitude agreements control is considered. A multi-group control architecture is designed, where one rigid body in each group is considered as the leader and the other rigid bodies are considered as followers. The designed intra-group cooperative controllers drive the attitude of followers in each group to that of the leader located in this group, and the constructed inter-group cooperative controllers drive all the attitudes of leaders to an anti-consensus configuration. The two kinds of distributed cooperative control algorithms accomplish the multiple attitude agreements task.
- Distributed robust finite-time attitude containment control law for multiple rigid bodies with uncertainties including parametric uncertainties, external disturbances, and actuator failures is firstly investigated. A sliding-mode surface is designed such that the state of each rigid body on this surface will converge to the desired value in a finite time. Based on this surface, two different types of distributed controllers are constructed under two different cases, respectively and both of them can drive the orientations of the followers into the convex hull formed by the orientations of leaders in a finite time.
- Data-driven distributed optimal consensus control problem for unknown MASs with input delays is addressed using adaptive dynamic programming method. A model reduction method is utilized to deal with input delays so as to convert inputdelayed MAS model into a delay-free one. Meanwhile, by establishing an equivalent relationship on the performance indices of the two MASs, optimal consensus control of inputdelayed MAS can be fully transformed to that of delay-free MAS. Then we develop a PI algorithm based on distributed asynchronous update mechanism to online learn the solutions of the coupled HJB equations. Distributed asynchronous scheme reduces computation and alleviates the occupation of network in comparison with the existing distributed synchronized PI method. The critic-actor neural networks can be utilized to approximate the value functions and control policies only using measurement data instead of the system identification methods, the approximation technique can avoid the approximation residual errors caused by system identification
- An optimal consensus control algorithm for unknown nonlinear MASs with control constraints is proposed. A new event-triggered adaptive dynamic programming schematic diagram which includes smart critic network and smart action network

is built. Under the event-triggered sampling mechanism, the weights of the critic-action networks are only updated at the triggering instants. This aperiodic updated scheme can save computation resource and reduce transmission load compared with the time-triggered adaptive dynamic programming method.

1.5 Book Organization

This book is organized as follows: Chap. 1 reviews the developments of distributed multiagent systems cooperation, and presents the analysis and synthesis of cooperative control method and communication framework of MASs. Part I composes of Chaps. 2 and 3, which focus on distributed cooperation problem of linear homogeneous and heterogeneous MASs. Part II consists of Chaps. 4 and 5, studying consensus problems of nonlinear MASs with communication delays and state saturations. Then the cooperation framework is applied to multiple rigid body systems in Part III including Chaps. 6 and 7. Part IV containing Chaps. 8 and 9 imposes ADP technique to address the optimal cooperative control problem.

In Chap. 2, the consensus problem for discrete-time MASs with communication delays is investigated using adaptive model-based event-triggered predictive control. Based on the event-triggered communication scheme together with the networked predictive control scheme, a novel distributed adaptive model-based event-triggered predictive control protocol is proposed to reduce the network communication burden and meanwhile compensate for the communication delays actively. By using the virtual interval segmentation technique and the time-delay modeling method together with defining a set of switching signals, the closed-loop systems are established in forms of switching time-delay systems. The problem of simultaneously deriving the controller gain matrix and the event-triggering parameter matrices is cast into standard linear matrix inequalities problem.

In Chap. 3, event-triggered tracking control algorithm for heterogeneous MASs with Markov communication delays is proposed. Distributed consensus control is firstly designed by employing the event-triggered communication technique, which can lead to a significant reduction of the information communication burden in the multi-agent network. Then, the mean square stability of the closed loop MASs is analyzed based on the Lyapunov functional method and the Kronecker product technique. Sufficient conditions are obtained to guarantee the consensus in terms of linear matrix inequalities.

In Chap. 4, time-varying formation tracking issue for a class of uncertain non-affine nonlinear MASs with communication delays is addressed. The agents are connected though a directed topology, and the non-affine problem is transformed into affine problem using the idea of active disturbance rejection control . The uncertainties of individual followers in the MAS are approximated via the extended state observers and are compensated in real time. Tracking differentiators are introduced to overcome the explosion of complexity in the derivatives of the desired signal and the virtual control variables. Local information from neighboring followers are utilized, and the Lyapunov-Krasovskii function is employed to analyse the stability theoretically.

In Chap. 5, continuous, intermittent and adaptive state feedback protocols are respectively proposed for achieving the constrained consensus of nonlinear MASs with relative state saturations. Utilizing the incidence matrix and the edge Laplacian, the consensus problem of nonlinear MASs with the relative state saturations can be cast into the stabilization problem of edge dynamics operating on the constrained set. Three types of protocols, namely continuous, intermittent and adaptive state feedback protocols are respectively proposed for achieving the constrained consensus, and meanwhile yielding consensus values. A consensus analysis is provided by virtue of state saturation theory, switched system theory, adaptive theory and Lyapunov stability theory. Output feedback protocol is also designed. Finally, the obtained results are applied to connectivity preservation for first order nonlinear MASs, despite the presence of limited communication range and input constraints.

In Chap. 6, distributed event-triggered cooperative attitude control of multiple groups of rigid bodies evolved on nonlinear manifold $SO(3)$ is investigated. The distributed event-triggered mechanism is introduced into the controller design to reduce the communication pressure. It is verified that, the orientations of rigid bodies in each group converge to a common value with zero angular velocities, and these common values in all groups achieve an anti-consensus configuration under the proposed intra- and inter-group cooperative control law.

In Chap. 7, distributed robust finite-time attitude containment control for multiple rigid bodies with uncertainties including parametric uncertainties, external disturbances and actuator failures is studied. Two types of distributed control laws are designed corresponding to two different cases respectively, and both of them can drive the orientations of the followers into the convex hull formed by the orientations of leaders in a finite time.

In Chap. 8, data-driven distributed optimal consensus control protocol for unknown MASs with input delays is proposed. The input-delayed MAS model is firstly converted into a delay-free form using a model reduction method. By establishing an equivalent relationship on the predesigned performance indices of the two MASs, optimal consensus control of input-delayed MAS can be fully transformed to that of delay-free MAS. Based on the coupled Hamilton-Jacobi equations and Bellman's optimality principle, optimal consensus control policies are derived for the transformed delay-free MAS. Then a policy iteration algorithm based on distributed asynchronous update mechanism is proposed to learn the coupled Hamilton-Jacobi-Bellman equations online. To perform the proposed data-driven adaptive dynamic programming algorithm, we adopt the measured data based critic-actor neural networks to approximate the value functions and the control policies, respectively.

In Chap. 9, event-triggered adaptive dynamic programming technique is used to consider optimal consensus control problem for unknown nonlinear MASs subjected to control constraints. To deal with the control constraints, we introduce nonquadratic energy consumption functions into performance indices and formulate the Hamilton-Jacobi-Bellman equations. Then, based on the Bellman's optimality principle, constrained optimal consensus control policies are designed from the HJB equations. In order to implement the event-triggered adaptive dynamic programming algorithm, the critic networks and action networks are developed to approximate the value functions and consensus control policies respectively based on the measurable system data. Under the event-triggered control framework, the weights of the critic networks and action networks are only updated at the triggering instants which are decided by the designed adaptive triggered conditions. The Lyapunov method is used to prove that the local neighbor consensus errors and the weight estimation errors of the critic networks and action networks are ultimately bounded.

References

1. A. Joshi, N. Ramakrishnan, E.N. Houstis, Multiagent systems to support networked scientific computing. IEEE Internet Comput. **2**(3), 69–83 (1998)
2. M.P. Singh, A.K. Chopra, The internet of things and multiagent systems: Decentralized intelligence in distributed computing, in *Proceeding of 37th International Conference on Distributed Computing Systems* (Atlanta, 2017), pp. 1738–1747
3. G. Weiss, *Multiagent Systems: A Modern Approach to Distributed Artificial Intelligence* (MIT press, 1999)
4. S. Kraus, A. Azaria, J. Fiosina, M. Greve, N. Hazon, L. Kolbe, T.-B. Lembcke, J.P. Müller, S. Schleibaum, M. Vollrath, AI for explaining decisions in multi-agent environments (2019). arXiv:1910.04404
5. E. Earon, T. Barfoot, G.D'Eleuterio, Development of a multiagent robotic system with application to space exploration, in *Proceeding of 2001 IEEE/ASME International Conference on Advanced Intelligent Mechatronics*, vol. 2 (Como, 2001), pp. 1267–1272
6. M. Colby, L. Yliniemi, K. Tumer, Autonomous multiagent space exploration with high-level human feedback. J. Aerosp. Inf. Syst. **13**(8), 301–315 (2016)
7. M. Hajduk, M. Sukop, M. Haun, *Cognitive Multi-agent Systems: Structures, Strategies and Applications to Mobile Robotics and Robosoccer*, vol. 138 (Springer, Berlin, 2018)
8. Q. Sun, Y. Tian, M. Diao, Cooperative localization algorithm based on hybrid topology architecture for multiple mobile robot system. IEEE Internet Things J. **5**(6), 4753–4763 (2018)
9. M.H. Moradi, S. Razini, S.M. Hosseinian, State of art of multiagent systems in power engineering: A review. Renew. Sustain. Energy Rev. **58**, 814–824 (2016)
10. A.S. Nair, T. Hossen, M. Campion, D.F. Selvaraj, N. Goveas, N. Kaabouch, P. Ranganathan, Multi-agent systems for resource allocation and scheduling in a smart grid. Technol. Econ. Smart Grids Sustain. Energy **3**(1), 15 (2018)
11. A. Farinelli, G. Grisetti, L. Iocchi, S.L. Cascio, D. Nardi, Design and evaluation of multi agent systems for rescue operations, in *Proceedings 2003 IEEE/RSJ International Conference on Intelligent Robots and Systems*, vol. 4 (Las Vegas, 2003), pp. 3138–3143
12. O. Malaschuk, A. Dyumin, Intelligent multi-agent system for rescue missions, in *Advanced Technologies in Robotics and Intelligent Systems* (Springer, Berlin, 2020), pp. 89–97
13. R. Olfati-Saber, J.A. Fax, R.M. Murray, Consensus and cooperation in networked multi-agent systems. Proc. IEEE **95**(1), 215–233 (2007)
14. L. Ding, Q.-L. Han, X. Ge, X.-M. Zhang, An overview of recent advances in event-triggered consensus of multiagent systems. IEEE Trans. Cybern. **48**(4), 1110–1123 (2017)
15. K.-K. Oh, M.-C. Park, H.-S. Ahn, A survey of multi-agent formation control. Automatica **53**, 424–440 (2015)
16. S. Knorn, Z. Chen, R.H. Middleton, Overview: Collective control of multiagent systems. IEEE Trans. Control Netw. Syst. **3**(4), 334–347 (2015)

17. J. Qin, Q. Ma, Y. Shi, L. Wang, Recent advances in consensus of multi-agent systems: A brief survey. IEEE Trans. Ind. Electron. **64**(6), 4972–4983 (2016)
18. Y. Cao, W. Yu, W. Ren, G. Chen, An overview of recent progress in the study of distributed multi-agent coordination. IEEE Trans. Ind. Inf. **9**(1), 427–438 (2012)
19. Y. Wang, E. Garcia, D. Casbeer, F. Zhang, *Cooperative Control of Multi-agent Systems: Theory and Applications* (Wiley, New York, 2017)
20. R.O. Saber, R.M. Murray, Consensus protocols for networks of dynamic agents, in *Proceedings of 2003 America Control Conference*, vol. 2 (Piscataway, 2003), pp. 951–956
21. R. Olfati-Saber, R.M. Murray, Consensus problems in networks of agents with switching topology and time-delays. IEEE Trans. Autom. Control **49**(9), 1520–1533 (2004)
22. C.-Q. Ma, J.-F. Zhang, Necessary and sufficient conditions for consensusability of linear multi-agent systems. IEEE Trans. Autom. Control **55**(5), 1263–1268 (2010)
23. W. Hou, M.Y. Fu, H. Zhang, Consensusability of linear multi-agent systems with time delay. Int. J. Robust Nonlinear Control **26**(12), 2529–2541 (2016)
24. L. Xu, N. Xiao, L. Xie, Consensusability of discrete-time linear multi-agent systems over analog fading networks. Automatica **71**, 292–299 (2016)
25. Z. Li, W. Ren, X. Liu, L. Xie, Distributed consensus of linear multi-agent systems with adaptive dynamic protocols. Automatica **49**(7), 1986–1995 (2013)
26. H. Zhang, D. Yue, X. Yin, S. Hu, C. xia Dou, Finite-time distributed event-triggered consensus control for multi-agent systems. Inf. Sci. **339**, 132–142 (2016)
27. L. Yu, J. Wang, Robust cooperative control for multi-agent systems via distributed output regulation. Syst. Control Lett. **62**(11), 1049–1056 (2013)
28. Y. Zhao, Y. Liu, G. Wen, W. Ren, G. Chen, Designing distributed specified-time consensus protocols for linear multiagent systems over directed graphs. IEEE Trans. Autom. Control **64**(7), 2945–2952 (2018)
29. C. Huang, X. Ye, Cooperative output regulation of heterogeneous multi-agent systems: An H_∞ criterion. IEEE Trans. Autom. Control **59**(1), 267–273 (2013)
30. X. Li, Y.C. Soh, L. Xie, F.L. Lewis, Cooperative output regulation of heterogeneous linear multi-agent networks via H_∞ performance allocation. IEEE Trans. Autom. Control **64**(2), 683–696 (2018)
31. H. Su, M. Chen, X. Wang et al., Semiglobal observer-based leader-following consensus with input saturation. IEEE Trans. Ind. Electron. **61**(6), 2842–2850 (2013)
32. Z. Pang, W. Luo, G. Liu, et al., Observer-based incremental predictive control of networked multi-agent systems with random delays and packet dropouts. IEEE Trans. Circuits Syst. II: Express Briefs (2020). https://doi.org/10.1109/TCSII.2020.2999126
33. Y. Wang, Y. Yuan, J. Liu, Finite-time leader-following output consensus for multi-agent systems via extended state observer. Automatica (2020). https://doi.org/10.1016/j.automatica.2020.109133
34. W. Dong, Consensus of high-order nonlinear continuous-time systems with uncertainty and limited communication data rate. IEEE Trans. Autom. Control **64**(5), 2100–2107 (2018)
35. M. Egerstedt, X. Hu, Formation constrained multi-agent control. IEEE Trans. Robotics Autom. **17**(6), 947–951 (2001)
36. Y. Liu, Z. Geng, Finite-time formation control for linear multi-agent systems: A motion planning approach. Syst. Control Lett. **85**, 54–60 (2015)
37. K.-K. Oh, H.-S. Ahn, Formation control and network localization via orientation alignment. IEEE Trans. Autom. Control **59**(2), 540–545 (2013)
38. D.V. Dimarogonas, K.H. Johansson, Stability analysis for multi-agent systems using the incidence matrix: Quantized communication and formation control. Automatica **46**(4), 695–700 (2010)
39. D. Li, S.S. Ge, W. He, G. Ma, L. Xie, Multilayer formation control of multi-agent systems. Automatica **109**, 108558 (2019)
40. M.A. Kamel, X. Yu, Y. Zhang, Formation control and coordination of multiple unmanned ground vehicles in normal and faulty situations: A review. Ann. Rev. Control **49**, 128–144 (2020)

41. F. Mehdifar, C.P. Bechlioulis, F. Hashemzadeh, M. Baradarannia, Prescribed performance distance-based formation control of multi-agent systems. Automatica **119**, 109086 (2020)
42. X. Sun, C.G. Cassandras, Optimal dynamic formation control of multi-agent systems in constrained environments. Automatica **73**, 169–179 (2016)
43. S. Li, J. Zhang, X. Li, F. Wang, X. Luo, X. Guan, Formation control of heterogeneous discrete-time nonlinear multi-agent systems with uncertainties. IEEE Trans. Ind. Electron. **64**(6), 4730–4740 (2017)
44. Q. Wang, M. Wu, Y. Huang, L. Wang, Formation control of heterogeneous multi-robot systems. IFAC Proc. Volumes **41**(2), 6596–6601 (2008)
45. S. Liu, L. Xie, H. Zhang, Containment control of multi-agent systems by exploiting the control inputs of neighbors. Int. J. Robust Nonlinear Control **24**(17), 2803–2818 (2014)
46. H. Liu, G. Xie, L. Wang, Necessary and sufficient conditions for containment control of networked multi-agent systems. Automatica **48**(7), 1415–1422 (2012)
47. J. Shao, L. Shi, W. Zhen, T. Huang, Containment control for heterogeneous multi-agent systems with asynchronous updates. Inf. Sci. **436**, 74–88 (2018); 6596–6601 (2008)
48. S. Weng, D. Yue, Z. Sun, L. Xiao, Distributed robust finite-time attitude containment control for multiple rigid bodies with uncertainties. Int. J. Robust Nonlinear Control **25**, 62561–2581 (2015)
49. Z. Meng, W. Ren, Y. Zheng, Distributed finite-time attitude containment control for multiple rigid bodies. Automatica **46**(12), 2092–2099 (2010)
50. Q. Zhou, W. Wang, H. Liang, B. Wang, Observer-based event-triggered fuzzy adaptive bipartite containment control of multi-agent systems with input quantization. IEEE Trans. Fuzzy Syst. (2019). https://doi.org/10.1109/TFUZZ.2019.2953573
51. S.J. Qin, Survey on data-driven industrial process monitoring and diagnosis. Ann. Rev. Control **36**(2), 220–234 (2012)
52. Z.-S. Hou, Z. Wang, From model-based control to data-driven control: Survey, classification and perspective. Inf. Sci. **235**, 3–35 (2013)
53. S. Yin, X. Li, H. Gao, O. Kaynak, Data-based techniques focused on modern industry: An overview. IEEE Trans. Ind. Electron. **62**(1), 657–667 (2014)
54. M. Benosman, Model-based vs data-driven adaptive control: An overview. Int. J. Adaptive Control Signal Process. **32**(5), 753–776 (2018)
55. H.-S. Ahn, Y. Chen, Iterative learning control for multi-agent formation, in *2009 ICROS-SICE International Joint Conference* (Fukuoka, 2009), pp. 3111–3116
56. Y. Liu, Y. Jia, An iterative learning approach to formation control of multi-agent systems. Syst. Control Lett. **61**(1), 148–154 (2012)
57. D. Meng, Y. Jia, J. Du, J. Zhang, On iterative learning algorithms for the formation control of nonlinear multi-agent systems. Automatica **50**(1), 291–295 (2014)
58. J. Li, J. Li, Adaptive iterative learning control for coordination of second-order multi-agent systems. Int. J. Robust Nonlinear Control **24**(18), 3282–3299 (2014)
59. X. Jin, Adaptive iterative learning control for high-order nonlinear multi-agent systems consensus tracking. Syst. Control Lett. **89**, 16–23 (2016)
60. X. Jin, Nonrepetitive leader-follower formation tracking for multiagent systems with los range and angle constraints using iterative learning control. IEEE Trans. Cybern. **49**(5), 1748–1758 (2018)
61. X. Bu, P. Zhu, Z. Hou, J. Liang, Finite-time consensus for linear multi-agent systems using data-driven terminal ilc. IEEE Trans. Circuits Syst. II, Express Briefs (2019)
62. X. Bu, J. Liang, Z. Hou, R. Chi, Data-driven terminal iterative learning consensus for nonlinear multiagent systems with output saturation. IEEE Trans. Neural Netw. Learn. Syst. (2020)
63. C. Hua, Y. Qiu, X. Guan, Event-triggered iterative learning containment control of model-free multiagent systems, IEEE Trans. Syst. Man Cybern. Syst. (2020)
64. K.G. Vamvoudakis, F.L. Lewis, G.R. Hudas, Multi-agent differential graphical games: Online adaptive learning solution for synchronization with optimality. Automatica **48**(8), 1598–1611 (2012)

65. M.I. Abouheaf, F.L. Lewis, K.G. Vamvoudakis, S. Haesaert, R. Babuska, Multi-agent discrete-time graphical games and reinforcement learning solutions. Automatica **50**(12), 3038–3053 (2014)
66. Q. Wei, D. Liu, F.L. Lewis, Optimal distributed synchronization control for continuous-time heterogeneous multi-agent differential graphical games. Inf. Sci. **317**(1), 96–113 (2015)
67. H. Zhang, J. Zhang, G.-H. Yang, Y. Luo, Leader-based optimal coordination control for the consensus problem of multiagent differential games via fuzzy adaptive dynamic programming. IEEE Trans. Fuzzy Syst. **23**(1), 152–163 (2015)
68. H. Zhang, H. Jiang, Y. Luo, G. Xiao, Data-driven optimal consensus control for discrete-time multi-agent systems with unknown dynamics using reinforcement learning method. IEEE Trans. Ind. Electron. **64**(5), 4091–4100 (2017)
69. H. Cai, F.L. Lewis, G. Hu, J. Huang, The adaptive distributed observer approach to the cooperative output regulation of linear multi-agent systems. Automatica **75**, 299–305 (2017)
70. H. Zhang, D. Yue, W. Zhao, S. Hu, C. Dou, Distributed optimal consensus control for multiagent systems with input delay. IEEE Trans. Cybern. **48**(6), 1747–1759 (2017)
71. W. Zhao, R. Li, H. Zhang, Leader-follower optimal coordination tracking control for multi-agent systems with unknown internal states. Neurocomputing **249**, 171–181 (2017)
72. F.A. Yaghmaie, F.L. Lewis, R. Su, Output regulation of linear heterogeneous multi-agent systems via output and state feedback. Automatica **67**, 157–164 (2016)
73. H. Modares, S.P. Nageshrao, G.A.D. Lopes, R. Babuška, F.L. Lewis, Optimal model-free output synchronization of heterogeneous systems using off-policy reinforcement learning. Automatica **71**, 334–341 (2016)
74. Q. Jiao, H. Modares, S. Xu, F.L. Lewis, K.G. Vamvoudakis, Multi-agent zero-sum differential graphical games for disturbance rejection in distributed control. Automatica **69**, 24–34 (2016)
75. W. Ren, R.W. Beard, E.M. Atkins, Information consensus in multivehicle cooperative control. IEEE Control Syst. Mag. **27**(2), 71–82 (2007)
76. X. Ge, Q.-L. Han, D. Ding, X.-M. Zhang, B. Ning, A survey on recent advances in distributed sampled-data cooperative control of multi-agent systems. Neurocomputing **275**, 1684–1701 (2018)
77. H. Chu, D. Yue, C. Dou, L. Chu, Adaptive pi control for consensus of multiagent systems with relative state saturation constraints. IEEE Trans. Cybern. (2019)
78. Y. Yang, X. Si, D. Yue, J. Tan, Time-varying formation tracking of uncertain non-affine non-linear multi-agent systems with communication delays. IEEE Trans. Ind. Electron. (2020)
79. K.J. Astrom, B.M. Bernhardsson, Comparison of riemann and lebesgue sampling for first order stochastic systems, in *Proceedings of the 41st IEEE Conference on Decision and Control*, vol. 2 (IEEE, 2002), pp. 2011–2016
80. P. Tabuada, Event-triggered real-time scheduling of stabilizing control tasks. IEEE Trans. Autom. Control **52**(9), 1680–1685 (2007)
81. W. Heemels, K.H. Johansson, P. Tabuada, An introduction to event-triggered and self-triggered control, in *Proceeding of 2012 IEEE Conference on Decision and Control*, vol. 2 (Las Vegas, 2012), pp. 3270–3285
82. D. Yue, E. Tian, Q.-L. Han, A delay system method for designing event-triggered controllers of networked control systems. IEEE Trans. Autom. Control **58**(2), 475–481 (2012)
83. D.V. Dimarogonas, E. Frazzoli, K.H. Johansson, Distributed event-triggered control for multi-agent systems. IEEE Trans. Autom. Control **57**(5), 1291–1297 (2011)
84. X. Yin, D. Yue, S. Hu, Distributed event-triggered control of discrete-time heterogeneous multi-agent systems. J. Franklin Inst. **350**(3), 651–669 (2013)
85. X. Ge, Q.-L. Han, Distributed formation control of networked multi-agent systems using a dynamic event-triggered communication mechanism. IEEE Trans. Ind. Electron. **64**(10), 8118–8127 (2017)
86. A. Amini, A. Asif, A. Mohammadi, Dynamic event-triggered formation control for multi-agent systems: A co-design optimization approach, in *Proceeding of 2020 American Control Conference* (Denver, 2020), pp. 707–712

87. Y. Yang, Y. Li, D. Yue, W. Yue, Adaptive event-triggered consensus control of a class of second-order nonlinear multiagent systems. IEEE Trans. Cybern. (2019) https://doi.org/10.1109/TCYB.2019.2900266
88. M. Cao, F. Xiao, L. Wang, Event-based second-order consensus control for multi-agent systems via synchronous periodic event detection. IEEE Trans. Autom. Control **60**(9), 2452–2457 (2015)

Part I
Distributed Cooperative Control of Linear Multi-agent Systems

Chapter 2
Distributed Event-Triggered Consensus Control of Homogeneous Multiagent Systems

This chapter investigates the consensus problem for discrete-time networked MASs, where the information is exchanged through a shared network with communication delays. Based on the event-triggered communication scheme together with the networked predictive control scheme, a novel distributed adaptive model-based event-triggered predictive control protocol is proposed to reduce the network communication burden and meanwhile compensate for the communication delays actively. By using the virtual interval segmentation technique and the time-delay modeling method together with defining a set of switching signals, the closed-loop systems are established in forms of switching time-delay systems. The problem of simultaneously deriving the controller gain matrix and the event-triggering parameter matrices is cast into standard linear matrix inequalities problem. A convincing simulation example is given to demonstrate the theoretical results.

The rest of this chapter is organized as follows. Section 2.1 presents the consensus problem to be solved in this chapter. The distributed model-based event-triggered predictive control (D-MB-ETPC) scheme for nonuniform delay case is proposed in Sect. 2.2. Uniform delay case is given in Sect. 2.3. Simulation results are presented to illustrate the effectiveness of the theoretical results in Sect. 2.4. Section 2.5 offers some conclusions.

2.1 Problem Formulation

Consider a discrete-time networked MASs consisting of N following agents indexed by $1, 2, \ldots N$ and a leader indexed by 0. Every agent and the leader are described by

$$x_i(t+1) = Ax_i(t) + Bu_i(t) \tag{2.1}$$

$$x_0(t+1) = Ax_0(t) \tag{2.2}$$

D. Yue et al., *Distributed Cooperative Control and Communication for Multi-agent Systems*, https://doi.org/10.1007/978-981-33-6718-0_2

where $x_i(t) \in R^n$ and $u_i(t) \in R^p$ denote the state vector and control input vector of the ith agent, respectively. $i \in S_N$, where $S_N \triangleq \{1, 2, \ldots, N\}$. A and B are constant system matrices with appropriate dimensions. x_0 represents the leader specified by the dynamics in (2.2).

Definition 2.1 [1] For MASs (2.1) and (2.2), protocol $u_i(t)$ is said to solve the leader-following consensus problem asymptotically, if the states of agents satisfy

$$\lim_{t\to\infty} \|x_i(t) - x_0(t)\| = 0, \forall i \in S_N.$$

Throughout this chapter, it is assumed that all agents communicate with neighboring agents via a shared network. Given the fact that the real communication network is often unreliable, communication delays between agents are inevitable exist. When there exist communication delays, the routine consensus protocol based on the traditional time-triggered scheme is as follows:

$$\begin{aligned} u_i(t) = -K\Bigg\{ &\sum_{j\in N_i} a_{ij}[x_i(t-\tau_i(t)) - x_j(t-\tau_{ij}(t))] + b_i(x_i(t-\tau_i(t)) \\ &- x_0(t-\tau_{i0}(t)))\Bigg\} \end{aligned}$$

where K is the feedback gain matrix to be designed, $\tau_i(t)$ is the inner time delay for agent i itself and $\tau_{ij}(t)$ means the transmission delay from agent j to agent i at time t, $j = 0, 1, 2, \ldots, N$. For generally, $b_i \geq 0$ is the coupling weight between agent i and the leader with the characteristic that $b_i > 0$ if agent i is connected to the leader, otherwise, $b_i = 0$. For the use of the following sections, we define $H = diag\{b_1, b_2 \ldots, b_N\}$ as the connection matrix for the leader with other agents.

Before proceeding, we first give some assumptions.

Assumption 2.1 The shared network for agents only exists in $S - C$ (sensor to controller) channel. Let $\underline{\tau}_i$ and $\overline{\tau}_i$ denote the lower-bound and upper-bound delays for agent i, respectively. $\underline{d} \triangleq \min_{i\in S_N}\{\underline{\tau}_i\}$ and $\overline{d} \triangleq \max_{i\in S_N}\{\overline{\tau}_i\}$. Then, all time delays $\tau_{ij}(t)$ satisfy $\underline{d} \leq \underline{\tau}_i \leq \tau_i(t), \tau_{ij}(t) \leq \overline{\tau}_i \leq \overline{d}, i, j \in S_N$.

Assumption 2.2 In order to compensate for the network delays actively, at each agent i's sensor side, there equips one predictive generator for the leader. Furthermore, at each agent's controller side, there equips predictive generators for agent i itself, for leader 0 and for all its other neighbors.

Assumption 2.3 Assume that the initial conditions $x_0(t_0)$ and $x_i(t_0)$, $i \in S_N$, are all sent to their neighbors without transmission delays.

Assumption 2.4 The directed graph has a directed spanning tree.

Assumption 2.5 In all the data transmission, each data is time-stamped.

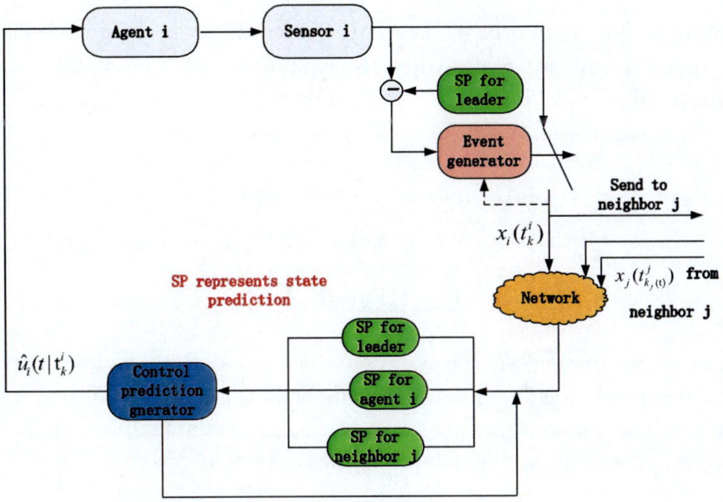

Fig. 2.1 Structure of the D-MB-ETPC for agent i

Remark 2.1 According to Assumptions 2.3 and 2.4, the leader's initial state $x_0(t_0)$ can be obtained by all agents at time t_0. This can be done as follows: first, $x_0(t_0)$ is sent by leader to its neighbors and leader's neighbors send it to all their own neighbors; then, the agents who received $x_0(t_0)$ will send it to their neighbors except for the ones who send it to themselves. By doing so, $x_0(t_0)$ can be obtained by all agents at time t_0 since the digraph has a directed spanning tree.

Remark 2.2 Taking into account Assumption 2.2 and Remark 2.1 and the dynamic for leader in (2.2), $x_0(t)(= Ax_0(t-1) = \cdots = A^{t-t_0}x_0(t_0))$ can be predicted at each agent's sensor side and controller side. In other words, after time t_0, the leader does not need to send its state information to its neighbors since all agents can predicate $x_0(t)$ by themselves just according to the obtained initial state $x_0(t_0)$. This conclusion will help us to design the distributed adaptive event-triggered communication scheme in the next section.

2.2 Distributed Consensus Control Design and Analysis

2.2.1 Distributed Adaptive Event-Triggered Strategy

In order to save the communication energy of the multi-agent network, here we adopt the distributed event-triggered communication scheme rather than the traditional time-triggered scheme. There exists an event detector at each agent i's sensor side, which is used to determine whether agent i sends its current state information or not to its neighbor through network, see Fig. 2.1. Under the event-triggered scheme, the kth event-triggered time and event-triggered state for agent i are denoted by t_k^i and $x_i(t_k^i)$, respectively, $k = 0, 1, 2, \ldots, i \in S_N$. For agent i, once $x_i(t_k^i)$ is sent out, the next event-triggered instant is determined by

$$t_{k+1}^i = t_k^i + \min\{l_k^i, q\} \tag{2.3}$$

$$\begin{aligned} l_k^i = \min_{l \in Z^+}\{l | [\delta_i(t_k^i + l) - \delta_i(t_k^i)]^T \Omega_i [\delta_i(t_k^i + l) - \delta_i(t_k^i)] \\ > \mu_i(t_k^i + l)\delta_i^T(t_k^i + l)\Omega_i \delta_i(t_k^i + l)\}, \end{aligned} \tag{2.4}$$

where q is a positive integer, Ω_i is a symmetric positive definite weighting matrix for agent i to be designed later, $\delta_i(t) = x_i(t) - \widehat{x}_0(t)$ ($\widehat{x}_0(t)$ is the predictive state for the leader and it will be shown that $\widehat{x}_0(t) = x_0(t)$), $\mu_i(\cdot)$ is the adaptive (time-varying) event-triggered parameter and its adaptive law is given as

$$\mu_i(t_k^i + l) - \mu_i(t_k^i + l - 1) = (\lambda - \mu_i(t_k^i + l))[\delta_i(t_k^i + l) - \delta_i(t_k^i)]^T \Omega_i [\delta_i(t_k^i + l) - \delta_i(t_k^i)] \tag{2.5}$$

where λ is any given positive constant. Based on Remark 2.2, it can be seen from the event-triggered scheme (2.3) that the next triggering instant t_{k+1}^i for agent i depends only on the local information. Any state after the latest transmitted state $x_i(t_k^i)$ satisfying $t_k^i + l < t_k^i + q$ and $h_i(\cdot, \cdot) \leq 0$ needs not to be transmitted, where the detailed expression of function $h_i(\cdot, \cdot) : R^n \times R^n \rightarrow R$, called as event-triggered function, is defined as follows

$$h_i(\cdot, \cdot) = [\delta_i(t_k^i + l) - \delta_i(t_k^i)]^T \Omega_i [\delta_i(t_k^i + l) - \delta_i(t_k^i)] - \mu_i(t_k^i + l)\delta_i^T(t_k^i + l)\Omega_i \delta_i(t_k^i + l). \tag{2.6}$$

In other words, for each agent i, only part of its states (i.e., only those states that satisfy $t_k^i + l \geq t_k^i + q$ or $h_i(\cdot, \cdot) > 0$) need to be transmitted. Consequently, the communications between agents can be reduced and the computation and communication resources can be saved for other important uses.

Remark 2.3 The main reasons for us to choose the adaptive law for event-triggered parameters as the form in (2.5) are that first it can help us to avoid the difficulties in choosing suitable initial parameters and second it is helpful for us to analyze the consensus performance of the investigated MASs under MB-D-ETPC scheme by using the switching Lyapunov functional method.

2.2.2 Distributed Consensus Control Design

In order to overcome the negative effects of transmission delays under above distributed adaptive event-triggered communication scheme, the networked model-based predictive control is adopted, we call this kind of control as D-MB-ETPC. The detailed realize process for this control are shown below.

We first take the initial time interval $[t_0, t_1^i + \tau_i(t_1^i))$ for instant. At the controller side of agent i, based on the obtained initial states $x_i(t_0)$ and $x_j(t_0)$, $j \in N_i$ and under Assumption 2.3 with Remarks 2.1 and 2.2, define $t_0^i = t_0$ and $\widehat{x}_i(t_0) = x_i(t_0)$, the future $t_1^i + \tau_i(t_1^i) - t_0 - 1$ step predictive states $\widehat{x}_i(t_0 + 1|t_0), \ldots, \widehat{x}_i(t_1^i|t_0), \ldots, \widehat{x}_i(t_1^i + \tau_i(t_1^i) - 1|t_0)$ and the predictive controllers $u_i(t_0|t_0)$, $\widehat{u}_i(t_0 + 1|t_0), \ldots, \widehat{u}_i(t_1^i + \tau_i(t_1^i) - 1|t_0)$ are calculated as follows:

$$u_i(t_0|t_0) = -K\{\sum_{j \in N_i} a_{ij}[x_i(t_0) - x_j(t_0)] + b_i(x_i(t_0) - x_0(t_0))\} \tag{2.7}$$

$$\widehat{x}_i(t_0 + 1|t_0) = Ax_i(t_0) + Bu_i(t_0|t_0) \tag{2.8}$$

$$\begin{aligned} \widehat{u}_i(t_0 + 1|t_0) = &-K\{\sum_{j \in N_i} a_{ij}[\widehat{x}_i(t_0 + 1|t_0) - \widehat{x}_j(t_0 + 1|t_0)] \\ &+ b_i(\widehat{x}_i(t_0 + 1|t_0) - \widehat{x}_0(t_0 + 1|t_0))\} \end{aligned} \tag{2.9}$$

$$\widehat{x}_i(t_0 + 2|t_0) = A\widehat{x}_i(t_0 + 1|t_0) + B\widehat{u}_i(t_0 + 1|t_0) \tag{2.10}$$

$$\vdots$$

$$\begin{aligned} \widehat{u}_i(t_1^i + \tau_i(t_1^i) - 2|t_0) = &-K\{\sum_{j \in N_i} a_{ij}[\widehat{x}_i(t_1^i + \tau_i(t_1^i) - 2|t_0) - \widehat{x}_j(t_1^i + \tau_i(t_1^i) \\ &- 2|t^j_{k_{ij}(t_1^i + \tau_i(t_1^i) - 2)})] + b_i(\widehat{x}_i(t_1^i + \tau_i(t_1^i) - 2|t_0) - \widehat{x}_0(t_1^i + \tau_i(t_1^i) - 2|t_0)\} \end{aligned} \tag{2.11}$$

$$\widehat{x}_i(t_1^i + \tau_i(t_1^i) - 1|t_0) = A\widehat{x}_i(t_1^i + \tau_i(t_1^i) - 2|t_0) + B\widehat{u}_i(t_1^i + \tau_i(t_1^i) - 2|t_0) \tag{2.12}$$

$$\begin{aligned} \widehat{u}_i(t_1^i + \tau_i(t_1^i) - 1|t_0) = &-K\{\sum_{j \in N_i} a_{ij}[\widehat{x}_i(t_1^i + \tau_i(t_1^i) - 1|t_0) - \widehat{x}_j(t_1^i + \tau_i(t_1^i) \\ &- 1|t^j_{k_{ij}(t_1^i + \tau_i(t_1^i) - 1)})] + b_i(\widehat{x}_i(t_1^i + \tau_i(t_1^i) - 1|t_0) - \widehat{x}_0(t_1^i + \tau_i(t_1^i) - 1|t_0)\} \end{aligned} \tag{2.13}$$

with

$$\begin{aligned} \widehat{x}_j(t|t^j_{k_{ij}(t)}) &= A\widehat{x}_j(t - 1|t^j_{k_{ij}(t)}) = A^2\widehat{x}_j(t - 2|t^j_{k_{ij}(t)}) \\ &= \cdots = A^{t - t^j_{k_{ij}(t)}} x_j(t^j_{k_{ij}(t)}) \end{aligned} \tag{2.14}$$

where

$$\begin{aligned} k_{ij}(t) &\triangleq \arg\min_m \{t - (t_m^j + \tau_{ij}(t_m^j)) | t \geq t_m^j + \tau_{ij}(t_m^j), m = 1, 2, \ldots\} \\ t &\in [t_0, t_1^i + \tau_i(t_1^i)) \cap [t^j_{k_{ij}(t)} + \tau_{ij}(t^j_{k_{ij}(t)}), t^j_{k_j(t)+1} + \tau_{ij}(t^j_{k_{ij}(t)+1})). \end{aligned}$$

After time t_0, the future $t_1^i + \tau_i(t_1^i) - t_0 - 1$ step state predictions for the leader are constructed as

$$\begin{aligned}\widehat{x}_0(t_0|t_0) &\triangleq x_0(t_0)\\ \widehat{x}_0(t_0+r|t_0) &= A\widehat{x}_0(t_0+r-1|t_0)\end{aligned} \tag{2.15}$$

where $r = 1, 2, \ldots, t_1^i + \tau_i(t_1^i) - t_0 - 1$.

Remark 2.4 Generally speaking, as stated in work [2], agent i has no sensor to detect the current or future control decision of its neighbors, i.e., $\widehat{u}_j(t^j_{k_{ij}(t)} + r|t^j_{k_{ij}(t)})$, $r = 0, 1, 2, \ldots$, thus in agent i's calculation for the predictive state of its neighbors j, $j \in N_i$ during the future $t_1^i + \tau_i(t_1^i) - t_0 - 1$ steps, it is assumed that $\widehat{u}_j(t^j_{k_{ij}(t)} + r|t^j_{k_{ij}(t)}) = 0$, $r = 0, 1, 2, \ldots$, as has been done in (2.14).

Since it is assumed that there doesn't exist network between the controller and the plant, for time $t \in [t_0, t_1^i + \tau_i(t_1^i))$, i.e., for time $t = t_0, t_0 + 1, \ldots, t_1^i + \tau_i(t_1^i) - 1$, the predictive controllers in (2.7), (2.9), (2.11) and (2.13) are acted on agent i's plant side successively without transmission delay. It should be pointed out that some of the predictive controllers in (2.7), (2.9), (2.11) and (2.13) need to be saved in a single data packet denoted by

$$U^i_{t_0} \triangleq \left\{\widehat{u}_i(t_1^i|t_0),\ \widehat{u}_i(t_1^i+1|t_0),\ \ldots,\ \widehat{u}_i(t_1^i+\tau_i(t_1^i)-1|t_0)\right\} \tag{2.16}$$

at agent i's controller side for the use of controller predictions during the next time interval $[t_1^i + \tau_i(t_1^i), t_2^i + \tau_i(t_2^i))$.

For the general time interval $[t_k^i + \tau_i(t_k^i), t_{k+1}^i + \tau_i(t_{k+1}^i))$, $k = 1, 2, \ldots$, at each agent i's controller side, based on the latest obtained event-triggered state information $x_i(t_k^i)$ of agent i and $x_j(t^j_{k_j(t)})$ of its neighbors, let $\widehat{x}_i(t_k^i|t_k^i) \triangleq x_i(t_k^i)$ and $M_k^i \triangleq t_{k+1}^i + \tau_i(t_{k+1}^i) - t_k^i - \tau_i(t_k^i) - 1$ ($M_k^i \le M$ with $M = \overline{d} + q - 1$), the future M_k^i step state predictions and controller predictions from time $t_k^i + \tau_i(t_k^i) + 1$ to time $t_k^i + \tau_i(t_k^i) + M_k^i$ are calculated as

$$\widehat{u}_i(t_k^i+s-1|t_k^i) = \widehat{u}_i(t_k^i+s-1|t_{k-1}^i) \in U^i_{t^i_{k-1}} \tag{2.17}$$

$$\begin{aligned}\widehat{x}_i(t_k^i+s|t_k^i) &= A\widehat{x}_i(t_k^i+s-1|t_k^i) + B\widehat{u}_i(t_k^i+s-1|t_k^i)\\ s &= 1, 2, \ldots \tau_i(t_k^i)\end{aligned} \tag{2.18}$$

and

$$\begin{aligned}\widehat{u}_i(t_k^i+s-1|t_k^i) = -K\Bigg\{&\sum_{j\in N_i} a_{ij}[\widehat{x}_i(t_k^i+s-1|t_k^i) - \widehat{x}_j(t_k^i+s\\ &-1|t^j_{k_{ij}(t_k^i+s-1)})] + b_i(\widehat{x}_i(t_k^i+s-1|t_k^i) - \widehat{x}_0(t_k^i+s-1))\Bigg\}\end{aligned} \tag{2.19}$$

$$\widehat{x}_i(t_k^i+s|t_k^i) = A\widehat{x}_i((t_k^i+s-1|t_k^i)) + B\widehat{u}_i(t_k^i+s-1|t_k^i) \tag{2.20}$$

$$\widehat{u}_i(t_k^i+s|t_k^i) = -K\Bigg\{\sum_{j\in N_i} a_{ij}[\widehat{x}_i(t_k^i+s|t_k^i) - \widehat{x}_j(t_k^i+s|t^j_{k_{ij}(t_k^i+s)})]$$

$$+ b_i(\widehat{x}_i(t_k^i + s|t_k^i) - \widehat{x}_0(t_k^i + s))\Big\} \tag{2.21}$$
$$s = \tau_i(t_k^i) + 1, \ldots, M_k^i$$

with

$$\begin{aligned}\widehat{x}_0(t) &= A\widehat{x}_0(t-1) = A^2\widehat{x}_0(t-2) \\ &= \cdots = A^t x_0(t_0) = x_0(t)\end{aligned} \tag{2.22}$$

and

$$\begin{aligned}\widehat{x}_j(t|t^j_{k_{ij}(t)}) &= A\widehat{x}_j(t-1|t^j_{k_{ij}(t)}) = A^2\widehat{x}_j(t-2|t^j_{k_{ij}(t)}) \\ &= \cdots = A^{t-t^j_{k_{ij}(t)}} x_j(t^j_{k_{ij}(t)})\end{aligned} \tag{2.23}$$

where

$$\begin{aligned}&k_{ij}(t) \triangleq \arg\min_m\{t - (t_m^j + \tau_{ij}(t_m^j))|t \geq t_m^j + \tau_{ij}(t_m^j), m = 1, 2, \ldots\} \\ &t \in [t_k^i + \tau_i(t_k^i), t_{k+1}^i + \tau_i(t_{k+1}^i)) \cap [t^j_{k_{ij}(t)} + \tau_{ij}(t^j_{k_{ij}(t)}), t^j_{k_{ij}(t)+1} + \tau_{ij}(t^j_{k_{ij}(t)+1})).\end{aligned} \tag{2.24}$$

Remark 2.5 From (2.23) and (2.24), it can be seen that agent i's calculation for the state of its neighbor j is based on the latest obtained event-triggered state $x_j(t^j_{k_{ij}(t)})$ of agent j.

Also, since it is assumed that there doesn't exist network between the controller and the plant, for time $t \in [t_k^i + \tau_i(t_k^i), t_{k+1}^i + \tau_i(t_{k+1}^i))$, i.e., for time $t = t_k^i + \tau_i(t_k^i), t_k^i + \tau_i(t_k^i) + 1, \ldots, t_{k+1}^i + \tau_i(t_{k+1}^i) - 1$, the predictive controllers in (2.17), (2.19) and (2.21) are acted on agent i's plant side successively without transmission delay. It should be pointed out that some of the predictive controllers in (2.17), (2.19) and (2.21) need to be saved in a single data packet denoted by

$$U^i_{t_k^i} \triangleq \left\{\widehat{u}_i(t_{k+1}^i|t_k^i), \widehat{u}_i(t_{k+1}^i + 1|t_k^i), \cdots, \widehat{u}_i(t_{k+1}^i + \tau_i(t_{k+1}^i) - 1|t_k^i)\right\} \tag{2.25}$$

at agent i's controller side for the use of controller predictions during the next time interval $[t_{k+1}^i + \tau_i(t_{k+1}^i), t_{k+2}^i + \tau_i(t_{k+2}^i))$.

Based on the above statement, we can design the control for agent i as follows:

$$u_i(t) = \widehat{u}_i(t|t_k^i) = -K\left\{\sum_{j\in N_i} a_{ij}[\widehat{x}_i(t|t_k^i) - \widehat{x}_j(t|t^j_{k_{ij}(t)})] + b_i(\widehat{x}_i(t|t_k^i) - x_0(t))\right\} \tag{2.26}$$
$$t \in [t_k^i + \tau_i(t_k^i), t_{k+1}^i + \tau_i(t_{k+1}^i)) \cap [t^j_{k_{ij}(t)} + \tau_{ij}(t^j_{k_{ij}(t)}), t^j_{k_{ij}(t)+1} + \tau_{ij}(t^j_{k_{ij}(t)+1})).$$

2.2.3 Modeling of The Closed-Loop Systems

The conclusion given in the following lemma is important for us to establish the closed-loop systems for systems (2.1) and (2.2) with control (2.26).

Lemma 2.1 *For any time $t \in [t_0, +\infty)$, the state $x_i(t)$ of system (2.1) for agent i is the same as its predictive state $\widehat{x}_i(t|t_k^i)$ constructed in (2.8), (2.10), (2.12), (2.18) and (2.20), that is*

$$x_i(t) = \widehat{x}_i(t|t_k^i)$$

for $t \in [t_k^i + \tau_i(t_k^i), t_{k+1}^i + \tau_i(t_{k+1}^i))$ *with* $\cup_{k=0}^{+\infty}[t_k^i + \tau_i(t_k^i), t_{k+1}^i + \tau_i(t_{k+1}^i)) = [t_0, +\infty)$.

Proof This lemma will be proved by using the mathematical induction method. For time $t = t_0$, the conclusion is true since we have defined $\widehat{x}_i(t_0) = x_i(t_0)$.

For $\forall m \in (t_0, +\infty)$, there exist a positive integer r and one event-triggered time interval with delay $[t_k^i + \tau_i(t_k^i), t_{k+1}^i + \tau_i(t_{k+1}^i))$, s.t. $m = (t_k^i + r) \in [t_k^i + \tau_i(t_k^i), t_{k+1}^i + \tau_i(t_{k+1}^i))$. Assume that $x_i(m) = \widehat{x}_i(m|t_k^i)$, now we just need to prove the following two cases: Case (1), $x_i(m+1) = \widehat{x}_i(m+1|t_k^i)$ if $(m+1) \in [t_k^i + \tau_i(t_k^i), t_{k+1}^i + \tau_i(t_{k+1}^i))$; Case (2), $x_i(m+1) = \widehat{x}_i(m+1|t_{k+1}^i)$ if $m+1 = t_{k+1}^i + \tau_i(t_{k+1}^i)$, that is when $m+1$ just belong to the time interval $[t_{k+1}^i + \tau_i(t_{k+1}^i), t_{k+2}^i + \tau_i(t_{k+2}^i))$.

For Case (1), from system (2.1), control (2.26) and the state prediction system (2.20), we have that

$$\begin{aligned} x_i(m+1) &= Ax_i(m) + Bu_i(m) \\ &= A\widehat{x}_i(m|t_k^i) + B\widehat{u}_i(m|t_k^i) \\ &= \widehat{x}_i(m+1|t_k^i). \end{aligned}$$

For Case (2), from (2.17) and (2.19), we have

$$\begin{aligned} \widehat{x}_i(t_{k+1}^i + 1|t_{k+1}^i) &= A\widehat{x}_i(t_{k+1}^i|t_{k+1}^i) + B\widehat{u}_i(t_{k+1}^i|t_k^i) \\ &= Ax_i(t_{k+1}^i) + B\widehat{u}_i(t_{k+1}^i|t_k^i) \\ &= x_i(t_{k+1}^i + 1) \\ \widehat{x}_i(t_{k+1}^i + 2|t_{k+1}^i) &= A\widehat{x}_i(t_{k+1}^i + 1|t_{k+1}^i) + B\widehat{u}_i(t_{k+1}^i + 1|t_k^i) \\ &= Ax_i(t_{k+1}^i + 1) + B\widehat{u}_i(t_{k+1}^i + 1|t_k^i) \\ &= x_i(t_{k+1}^i + 2) \\ &\ \ \vdots \\ \widehat{x}_i(m+1|t_{k+1}^i) &= \widehat{x}_i(t_{k+1}^i + \tau_i(t_{k+1}^i)|t_{k+1}^i) \\ &= A\widehat{x}_i(t_{k+1}^i + \tau_i(t_{k+1}^i) - 1|t_k^i) + B\widehat{u}_i(t_{k+1}^i + \tau_i(t_{k+1}^i) - 1|t_k^i) \\ &= A\widehat{x}_i(m|t_k^i) + B\widehat{u}_i(m|t_k^i) \\ &= Ax_i(m) + B\widehat{u}_i(m|t_k^i) \\ &= x_i(m+1). \end{aligned}$$

Combing the proof for Case (1) and Case (2), the conclusion in this lemma can be guaranteed.

Taking into account (2.22), (2.23) and Lemma 2.1, the control in (2.26) can be rewritten as

$$u_i(t) = -K\Big\{\sum_{j\in N_i} a_{ij}[x_i(t) - A^{t-t^j_{k_{ij}(t)}} x_j(t^j_{k_{ij}(t)})] + b_i(x_i(t) - x_0(t))\Big\} \tag{2.27}$$

$$t \in [t^i_k + \tau_i(t^i_k), t^i_{k+1} + \tau_i(t^i_{k+1})) \cap [t^j_{k_{ij}(t)} + \tau_{ij}(t^j_{k_{ij}(t)}), t^j_{k_{ij}(t)+1} + \tau_{ij}(t^j_{k_{ij}(t)+1})).$$

The closed-loop system of systems (2.1) can be further obtained as follows:

$$x_i(t+1) = Ax_i(t) - BK\Big\{\sum_{j\in N_i} a_{ij}[x_i(t) - A^{t-t^j_{k_{ij}(t)}} x_j(t^j_{k_{ij}(t)})] + b_i(x_i(t) - x_0(t))\Big\} \tag{2.28}$$

$$t \in [t^i_k + \tau_i(t^i_k), t^i_{k+1} + \tau_i(t^i_{k+1})) \cap [t^j_{k_{ij}(t)} + \tau_{ij}(t^j_{k_{ij}(t)}), t^j_{k_{ij}(t)+1} + \tau_{ij}(t^j_{k_{ij}(t)+1})).$$

Noting that $\delta_i(t) = x_i(t) - x_0(t)$ and according to (2.28) and (2.2), we can obtain the closed-loop system for agent i as

$$\begin{aligned}\delta_i(t+1) &= A\delta_i(t) - BK\Big\{\sum_{j\in N_i} a_{ij}[x_i(t) - x_0(t) + x_0(t) - A^{t-t^j_{k_{ij}(t)}} x_j(t^j_{k_{ij}(t)})] + b_i\delta_i(t)\Big\}\\
&= A\delta_i(t) - BK\Big\{\sum_{j\in N_i} a_{ij}[\delta_i(t) + A^{t-t^j_{k_{ij}(t)}} x_0(t^j_{k_{ij}(t)}) - A^{t-t^j_{k_{ij}(t)}} x_j(t^j_{k_{ij}(t)})] + b_i\delta_i(t)\Big\}\\
&= A\delta_i(t) - BK\Big\{\sum_{j\in N_i} a_{ij}[\delta_i(t) - A^{t-t^j_{k_{ij}(t)}} \delta_j(t^j_{k_{ij}(t)})] + b_i\delta_i(t)\Big\}. \end{aligned} \tag{2.29}$$

For system (2.29), to achieve a detailed timing analysis, we use the virtual interval segmentation technique and time-delay modeling method [3, 4], let

$$\Theta = [t^j_{k_{ij}(t)} + \tau_{ij}(t^j_{k_{ij}(t)}), t^j_{k_{ij}(t)+1} + \tau_{ij}(t^j_{k_{ij}(t)+1})) = \cup\Theta_l$$
$$\Theta_l = [t^j_{k_{ij}(t)} + l + s^j_l, t^j_{k_{ij}(t)} + l + 1 + s^j_{l+1})$$

with $l = 0, 1, \ldots, t^j_{k_{ij}(t)+1} - t^j_{k_{ij}(t)} - 1$. If $l = t^j_{k_{ij}(t)+1} - t^j_{k_{ij}(t)} - 1$, then $s^i_{l+1} = \tau_{ij}(t^j_{k_{ij}(t)+1})$, otherwise, $s^i_l = \tau_{ij}(t^j_{k_{ij}(t)})$.

Define

$$\eta_{ij}(t) = t - t^j_{k_{ij}(t)} - l \tag{2.30}$$

$$e_{ij}(t - \eta_{ij}(t)) = \delta_j(t^j_{k_{ij}(t)}) - \delta_j(t^j_{k_{ij}(t)} + l) \tag{2.31}$$
$$t \in \Theta_l.$$

From (2.30), it can be easily obtained that

$$\underline{\tau}_j \leq \tau_{ij}(t^j_{k_{ij}(t)}) \leq \eta_{ij}(t) \leq \overline{d} + 1 \triangleq \eta_M.$$

According to (2.30)–(2.31), the definition of $k_{ij}(t)$ in (2.24) and the event-triggered conditions in (2.3) and (2.4), we can obtain that

$$\delta_j(t^j_{k_{ij}(t)}) = \delta_j(t - \eta_{ij}(t)) + e_{ij}(t - \eta_{ij}(t))$$

Then system (2.29) can be equivalently written as

$$\begin{aligned}\delta_i(t+1) = A\delta_i(t) - BK\Big\{ &\sum_{j\in N_i} a_{ij}[\delta_i(t) - A^{t-t^j_{k_{ij}(t)}}\delta_j(t - \eta_{ij}(t)) \\ &- A^{t-t^j_{k_{ij}(t)}} e_{ij}(t - \eta_{ij}(t))] + b_i\delta_i(t)\Big\}\end{aligned} \tag{2.32}$$

$$t \in [t^i_k + \tau_i(t^i_k), t^i_{k+1} + \tau_i(t^i_{k+1})) \cap [t^j_{k_{ij}(t)} + \tau_{ij}(t^j_{k_{ij}(t)}), t^j_{k_{ij}(t)+1} + \tau_{ij}(t^j_{k_{ij}(t)+1})).$$

Define the augmented variables as

$$\begin{aligned}
\delta(t) &= \left[\, \delta_1^T(t)\ \delta_2^T(t)\ \ldots\ \delta_N^T(t) \,\right]^T \\
e_i(t - \eta_i(t)) &= \left[\, e_{i1}^T(t - \eta_{i1}(t))\ e_{i2}^T(t - \eta_{i2}(t))\ \ldots\ e_{iN}^T(t - \eta_{iN}(t)) \,\right]^T \\
e(t - \eta(t)) &= \left[\, e_1^T(t - \eta_1(t))\ e_2^T(t - \eta_2(t))\ \ldots\ e_N^T(t - \eta_N(t)) \,\right]^T \\
\delta(t - d_m(t)) &= \left[\, \delta_1^T(t - d_m(t))\ \delta_2^T(t - d_m(t))\ \cdots\ \delta_N^T(t - d_m(t)) \,\right]^T
\end{aligned}$$

$m = 1, 2, \ldots, s$, where s is the number of the different delay values in $\{\eta_{ij}(t), 1 \leq i, j \leq N\}$, we can further obtain the following augmented closed-loop system (2.32) for the investigated NMASs:

$$\begin{aligned}\delta(t+1) = (I_N \otimes A - D \otimes BK - H \otimes BK)\delta(t) + \sum_{m=1}^{s}\sum_{uv\in\alpha_m} C_{muv} \\ \otimes BKA^{\sigma_{uv}(t)}\delta(t - d_m(t)) + \widetilde{A} \otimes BK\widetilde{A}^{\sigma(t)} e(t - \eta(t)).\end{aligned} \tag{2.33}$$

where $\alpha_m = \{uv : \eta_{uv}(t) = d_m(t)\}$, $\sigma_{uv}(t) = t - t^v_{k_{uv}(t)}$ is the introduced switching signal for agent v in agent u's control calculation at time t. It should be noted that $\sigma_{uv}(t) \in \varpi \triangleq \{1, 2, \ldots, p\}$, $p = q + \overline{d}$. $d_m(t) \in \{\eta_{uv}(t), 1 \leq u, v \leq N\}$. Furthermore,

$$C_{muv}(ij) = \begin{cases} a_{ij} & uv \in \alpha_m, i = u, j = v, i \neq j \\ 0 & \text{otherwise} \end{cases}$$

and

$$
\begin{aligned}
\widetilde{\mathcal{A}} &= diag\left\{\beta_1, \beta_2, \ldots, \beta_n\right\}, \\
\widetilde{A}^{\sigma(t)} &= diag\left\{\widetilde{A}_1^{\sigma_1(t)}, \widetilde{A}_2^{\sigma_2(t)}, \ldots, \widetilde{A}_N^{\sigma_N(t)}\right\}, \\
\widetilde{A}_i^{\sigma_i(t)} &= diag\left\{A^{\sigma_{i1}(t)}, \ldots, A^{\sigma_{ii-1}(t)}, 0, A^{\sigma_{ii+1}(t)}, \ldots, A^{\sigma_{iN}(t)}\right\},
\end{aligned}
$$

$i = 1, 2, \ldots, N$, where β_l is the lth row of the adjacency matrix $\mathcal{A}$, $l = 1, 2, \ldots, n$. To simplify the closed-loop system (2.33), we define the following matrices as

$$
\begin{aligned}
H_m^{\sigma(t)} &= \sum_{uv \in \alpha_m} C_{mvu} \otimes BKA^{\sigma_{uv}(t)}, \\
H^{\sigma(t)} &= \widetilde{\mathcal{A}} \otimes BK\widetilde{A}^{\sigma(t)},
\end{aligned}
$$

$\sigma(t) = [\sigma_1(t), \sigma_2(t), \ldots, \sigma_N(t)]$ with $\sigma_i(t) = [\sigma_{i1}(t), \sigma_{i2}(t), \ldots, \sigma_{iN}(t)]$. Then systems (2.33) can be written as

$$
\begin{aligned}
\delta(t+1) = {} & (I_N \otimes A - D \otimes BK - H \otimes BK)\delta(t) \\
& + \sum_{m=1}^{s} H_m^{\sigma(t)}\delta(t - d_m(t)) + H^{\sigma(t)} e(t - \eta(t)).
\end{aligned} \tag{2.34}
$$

2.2.4 Consensus Analysis

For the analysis use in the following, we first give the following Lemma 2.2.

Lemma 2.2 *If all communication delays are zero, let $e_i(t) = \delta_i(t) - \delta_i(t_k^i)$ for $t \in [t_k^i, t_{k+1}^i)$. Then, for given initial values $\mu_i(0) \in (0, \lambda]$, we have that*

$$e_i^T(t)\Omega_i e_i(t) \le \mu_i(t)\delta_i^T(t)\Omega_i\delta_i(t)$$

and furthermore

$$\mu_i(0) \le \mu_i(t) \le \lambda$$

$i \in S_N$, $t \in [t_0, +\infty) = \cup_{k=0}^{\infty}[t_k^i, t_{k+1}^i)$ with $t_0 = 0$.

Proof From the event-triggering function (2.6), the following inequality holds for any $t \in \{1, 2, \ldots\}$

$$e_i^T(t)\Omega_i e_i(t) \le \mu_i(t)\delta_i^T(t)\Omega_i\delta_i(t).$$

In the following, we will prove this lemma by using the mathematical induction method.

For $t = 1$, from (2.5) we can obtain that

$$\mu_i(1) - \mu_i(0) = (\lambda - \mu_i(1))e_i^T(0)\Omega_i e_i(0),$$

tanking into account that $\mu_i(0) \in (0, \lambda]$ and $e_i^T(t)\Omega_i e_i(t) \geq 0$, we have

$$\begin{aligned}\mu_i(1) &= \frac{\mu_i(0) + \lambda e_i^T(0)\Omega_i e_i(0)}{1 + e_i^T(0)\Omega_i e_i(0)} \\ &\leq \frac{\lambda + \lambda e_i^T(0)\Omega_i e_i(0)}{1 + e_i^T(0)\Omega_i e_i(0)} \\ &= \lambda \end{aligned} \tag{2.35}$$

and

$$\begin{aligned}\mu_i(1) &= \frac{\mu_i(0) + \lambda e_i^T(0)\Omega_i e_i(0)}{1 + e_i^T(0)\Omega_i e_i(0)} \\ &= \frac{(1 + e_i^T(0)\Omega_i e_i(0))\mu_i(0) + (\lambda - \mu_i(0))e_i^T(0)\Omega_i e_i(0)}{1 + e_i^T(0)\Omega_i e_i(0)} \\ &\geq \mu_i(0). \end{aligned} \tag{2.36}$$

For $\forall t \in \{2, 3, \ldots\}$, it is assumed that $\mu_i(0) \leq \mu_i(t) \leq \lambda$, now we just need to prove that $\mu_i(0) \leq \mu_i(t+1) \leq \lambda$. Also, due to the relationship (2.5), we can derive that

$$\mu_i(t+1) = \frac{\mu_i(t) + \lambda e_i^T(t)\Omega_i e_i(t)}{1 + e_i^T(t)\Omega_i e_i(t)} \tag{2.37}$$

Similar to (2.35) and (2.36), the above equality (2.37) implies that $\mu_i(0) \leq \mu_i(t) \leq \mu_i(t+1) \leq \lambda$. This completes the proof.

Remark 2.6 According to the event-triggering conditions (2.3) and (2.4) and Lemma 2.2, we can easily obtain that

$$\begin{aligned}e^T(t-\eta(t))\widetilde{\Omega}e(t-\eta(t)) &\leq \sum_{m=1}^{s}\sum_{uv\in\alpha_m} \lambda\delta_v^T(t-d_m(t))\Omega_v\delta_v(t-d_m(t)) \\ &= \sum_{m=1}^{s} \delta^T(t-d_m(t))\lambda\widehat{\Omega}_m\delta(t-d_m(t)) \end{aligned} \tag{2.38}$$

where $\widetilde{\Omega} = I_N \otimes \Omega$ with

$$\Omega = diag\left\{\Omega_1, \Omega_2, \cdots, \Omega_N\right\}$$

and

$$\widehat{\Omega}_m = diag\left\{\sum\nolimits_{i1\in\alpha_m}\Omega_1, \sum\nolimits_{i2\in\alpha_m}\Omega_2, \cdots, \sum\nolimits_{iN\in\alpha_m}\Omega_N\right\}.$$

In the following, we will derive sufficient conditions on the stability of systems (2.34) under the D-MB-ETPC. The design for the event-triggering parameters Ω_i in

(2.6) and the feedback gain matrix K in (2.27) will be carried out in terms of LMI techniques.

Theorem 2.1 *Under the local event-triggered control laws (2.27) with the event-triggering conditions (2.3)–(2.5), for given system matrices A, B and feedback gain matrix K, $\lambda > 0$, $\overline{d}_m \leq \eta_M$, the group of the follower agents (2.1) achieve leader-following consensus with time-varying communication delays, if there exist positive definite matrices $P_r > 0$, $Q_{mr} > 0$, $R_{mr} > 0$, $Z_m > 0$, $\Omega_i > 0$, $i \in S_N$, matrices X_m and Y_m with suitable dimensions such that the following inequalities are satisfied for all $r = [r_{11}, r_{12}, \ldots, r_{NN}]$, $r_{ij} \in \varpi$ and $m = 1, 2, \ldots, s$*

$$\begin{bmatrix} \Pi_r + \Psi + \Psi^T + \sum_{m=1}^{s} \overline{d}_m Z_m & * & * & * & * \\ F_1 & -P_r & * & * & * \\ \Xi_1 & 0 & -R_{1r} & * & * \\ \vdots & 0 & 0 & \ddots & * \\ \Xi_s & 0 & 0 & 0 & -R_{sr} \end{bmatrix} < 0, \tag{2.39}$$

$$\begin{bmatrix} Z_m & X_m \\ X_m^T & R_{mr} \end{bmatrix} \geq 0, \tag{2.40}$$

$$\begin{bmatrix} Z_m & Y_m \\ Y_m^T & R_{mr} \end{bmatrix} \geq 0, \tag{2.41}$$

where

$$\Pi_r = \begin{bmatrix} \sum_{m=1}^{s} Q_{mr} - P_r & * & * & * & * & * & * & * \\ 0 & \lambda\widehat{\Omega}_1 & * & * & * & * & * & * \\ 0 & 0 & \ddots & * & * & * & * & * \\ 0 & 0 & 0 & \lambda\widehat{\Omega}_s & * & * & * & * \\ 0 & 0 & 0 & 0 & -Q_{1r} & * & * & * \\ 0 & 0 & 0 & 0 & 0 & \ddots & * & * \\ 0 & 0 & 0 & 0 & 0 & 0 & -Q_{sr} & * \\ 0 & 0 & 0 & 0 & 0 & 0 & 0 & -\widetilde{\Omega} \end{bmatrix} \tag{2.42}$$

$$F_1 = \left[P_r [I_N \otimes A - (D + H) \otimes BK] \; P_r H_1^r \cdots P_r H_s^r \; 0 \cdots 0 \; P_r H^r \right] \tag{2.43}$$

$$\Xi_m = \sqrt{\overline{d}_m} \left[R_{mr}(I_N \otimes (A - I_n) - (D + H) \otimes BK) \; R_{mr} H_1^r \cdots R_{mr} H_s^r \right.$$
$$\left. 0 \cdots 0 \; R_{mr} H^r \right] \tag{2.44}$$

$$\tag{2.45}$$

$$\Psi = \left[\sum_{m=1}^{s} X_m \; Y_1 - X_1 \cdots Y_s - X_s \; -Y_1 \cdots -Y_s \; 0 \right]. \tag{2.46}$$

Proof Assume that $\sigma(t) = [r_{11}, r_{12}, \ldots, r_{NN}] \triangleq r$ at time t, $r_{ij} \in \varpi$, construct the following switching Lyapunov functional candidate as

$$V(t, \delta(t)) = V_1(t, \delta(t)) + V_2(t, \delta(t)) + V_3(t, \delta(t))$$

where

$$\begin{aligned}
V_1(t, \delta(t)) &= \delta^T(t) P_r \delta(t) \\
V_2(t, \delta(t)) &= \sum_{m=1}^{s} \sum_{\theta=t-\bar{d}_m}^{t-1} \delta^T(\theta) Q_{mr} \delta(\theta) \\
V_3(t, \delta(t)) &= \sum_{m=1}^{s} \sum_{j=-\bar{d}_m}^{-1} \sum_{\beta=t+j}^{t-1} \varsigma^T(\theta) R_{mr} \varsigma(\theta) \\
\varsigma(\theta) &= \delta(\theta+1) - \delta(\theta)
\end{aligned}$$

where $P_r, Q_{mr}, R_{mr} \in R^{nN}$ are positive definite matrices to be determined, $\bar{d}_m = \sup_t d_m(t)$. For any given time t, calculating the differences of $V_i(t, \delta(t))$ $(i = 1, 2, 3)$ along the trajectories of (2.34), we have

$$\begin{aligned}
\Delta V_1(t, \delta(t)) = {} & [(I_N \otimes A - D \otimes BK - H \otimes BK)\delta(t) + \sum_{m=1}^{s} H_m^{\sigma(t)} \delta(t - d_m(t)) \\
& + H^{\sigma(t)} e(t - \eta(t))]^T P_r [(I_N \otimes A - D \otimes BK - H \otimes BK)\delta(t) \\
& + \sum_{m=1}^{s} H_m^{\sigma(t)} \delta(t - d_m(t)) + H^{\sigma(t)} e(t - \eta(t))] - \delta^T(t) P_r \delta(t)
\end{aligned}$$

$$\begin{aligned}
\Delta V_2(t, \delta(t)) &= \sum_{m=1}^{s} \sum_{\theta=t+1-\bar{d}_m}^{t} \delta^T(\theta) Q_{mr} \delta(\theta) - \sum_{m=1}^{s} \sum_{\theta=t-\bar{d}_m}^{t-1} \delta^T(\theta) Q_{mr} \delta(\theta) \\
&= \sum_{m=1}^{s} [\delta^T(t) Q_{mr} \delta(t) - \delta^T(t - \bar{d}_m) Q_{mr} \delta(t - \bar{d}_m)] \\
\Delta V_3(t, \delta(t)) &= \sum_{m=1}^{s} \sum_{j=-\bar{d}_m}^{-1} \sum_{\beta=t+1+j}^{t} \varsigma^T(\theta) R_{mr} \varsigma(\theta) - \sum_{m=1}^{s} \sum_{j=-\bar{d}_m}^{-1} \sum_{\beta=t+j}^{t-1} \varsigma^T(\theta) R_{mr} \varsigma(\theta) \\
&= \sum_{m=1}^{s} \bar{d}_m \varsigma^T(t) R_{mr} \varsigma(t) - \sum_{m=1}^{s} \sum_{j=t-\bar{d}_m}^{t-1} \varsigma^T(j) R_{mr} \varsigma(j).
\end{aligned}$$

By using the free weighting matrix method and taking (2.38) in Remark 2.6 into account, we have that

$$
\begin{aligned}
\Delta V(t,\delta(t)) \le\ & \Delta V_1(t,\delta(t)) + \Delta V_2(t,\delta(t)) + \Delta V_3(t,\delta(t)) \\
& +2\sum_{m=1}^{s} \xi^T(t) X_m \left[\delta(t) - \delta(t-d_m(t)) - \sum_{j=t-d_m(t)}^{t-1} \varsigma(j) \right] \\
& +2\sum_{m=1}^{s} \xi^T(t) Y_m \left[\delta(t-d_m(t)) - \delta(t-\overline{d}_m) - \sum_{j=t-\overline{d}_m}^{t-d_m(t)-1} \varsigma(j) \right] \\
& +\sum_{m=1}^{s} \left[\overline{d}_m \xi^T(t) Z_m \xi(t) - \sum_{j=t-\overline{d}_m}^{t-1} \xi^T(t) Z_m \xi(t) \right] \\
& +\sum_{m=1}^{s} \delta^T(t-d_m(t)) \lambda \widehat{\Omega}_m \delta(t-d_m(t)) - e^T(t-\eta(t)) \widetilde{\Omega} e(t-\eta(t))
\end{aligned}
$$

with X_m and Y_m are matrices with suitable dimensions and $Z_m > 0$ and

$$
\xi^T(t) = \left[\, \delta^T(t)\ \delta^T(t-d_1(t))\ \cdots\ \delta^T(t-d_s(t))\ \delta^T(t-\overline{d}_1)\ \cdots\ \delta^T(t-\overline{d}_s)\ e^T(t-\eta(t)) \,\right],
$$

it yields that

$$
\begin{aligned}
\Delta V(t,\delta(t)) \le\ & \xi^T(t) \left\{ \Pi_r + F_1^T P_r^{-1} F_1 + \Psi + \Psi^T + \sum_{m=1}^{s} \overline{d}_m Z_m + \sum_{m=1}^{s} \Xi_m^T R_{mr}^{-1} \Xi_m \right\} \xi(t) \\
& - \sum_{m=1}^{s} \sum_{j=t-d_m(t)}^{t-1} \left[\xi^T(t)\ \varsigma^T(j) \right] \begin{bmatrix} Z_m & X_m \\ X_m^T & R_{mr} \end{bmatrix} \begin{bmatrix} \xi(t) \\ \varsigma(j) \end{bmatrix} \\
& - \sum_{m=1}^{s} \sum_{j=t-\overline{d}_m}^{t-d_m(t)-1} \left[\xi^T(t)\ \varsigma^T(j) \right] \begin{bmatrix} Z_m & Y_m \\ Y_m^T & R_{mr} \end{bmatrix} \begin{bmatrix} \xi(t) \\ \varsigma(j) \end{bmatrix}
\end{aligned} \tag{2.47}
$$

To solve the consensus problem by using the switching Lyapunov functional method, the right-hand side of (2.47) needs to be less than zero. This can be guaranteed by letting

$$
\Pi_r + F_1^T P_r^{-1} F_1 + \Psi + \Psi^T + \sum_{m=1}^{s} \overline{d}_m Z_m + \sum_{m=1}^{s} \Xi_m^T R_{mr}^{-1} \Xi_m < 0 \tag{2.48}
$$

and

$$
\begin{bmatrix} Z_m & X_m \\ X_m^T & R_{mr} \end{bmatrix} \ge 0
$$

$$
\begin{bmatrix} Z_m & Y_m \\ Y_m^T & R_{mr} \end{bmatrix} \ge 0
$$

for $\forall m = 1, 2, \ldots, s$. By using Schur complement, the condition in (2.48) is equivalent to

$$\begin{bmatrix} \Pi_r + \Psi + \Psi^T + \sum_{m=1}^{s} \bar{d}_m Z_m & * & * & * & * \\ \digamma_1 & -P_r & * & * & * \\ \Xi_1 & 0 & -R_{1r} & * & * \\ \vdots & 0 & 0 & \ddots & * \\ \Xi_s & 0 & 0 & 0 & -R_{sr} \end{bmatrix} < 0$$

with $\Pi_r, \Psi, \digamma_1, \Xi_m, m = 1, 2, \ldots, s$ given in (2.42)–(2.46).

Remark 2.7 The sufficient conditions in Theorem 2.1 are not in the form of LMIs since the nonlinear terms in $\digamma_1$ and Ξ_m and so can't be solved by using the MATLAB Toolbox. In the following theorem, we will derive the sufficient conditions in forms of LMIs to guarantee the consensus. Meanwhile, the feedback gain matrix K and the triggering matrices Ω_i can also be co-designed in the following theorem.

Theorem 2.2 *Under the local event-triggered control laws (2.27) with the event-triggering conditions (2.3)–(2.5), for given system matrices* $A, B, \lambda > 0, \rho > 0, \rho_{1m} > 0, \rho_{2m} > 0$, *the group of the follower agents (2.1) achieve leader-following consensus with time-varying communication delays, if there exist positive definite matrices* $\widetilde{P}_r > 0, Q_{mr} > 0, \widetilde{R}_{mr} > 0, Z_m > 0, \Omega_i > 0, i \in S_N$, *matrices* X_m *and* Y_m *with suitable dimensions such that the following LMIs are satisfied for all* $r = [r_{11}, r_{12}, \ldots, r_{NN}], r_{ij} \in \varpi$ *and* $m = 1, 2, \ldots, s$

$$\begin{bmatrix} \widetilde{\Pi}_r + \Psi + \Psi^T + \sum_{m=1}^{s} \bar{d}_m Z_m & * & * & * & * \\ \widetilde{\digamma}_1 & -\widetilde{P}_r & * & * & * \\ \widetilde{\Xi}_1 & 0 & -\widetilde{R}_{1r} & * & * \\ \vdots & 0 & 0 & \ddots & * \\ \widetilde{\Xi}_s & 0 & 0 & 0 & -\widetilde{R}_{sr} \end{bmatrix} < 0, \tag{2.49}$$

$$\begin{bmatrix} Z_m & X_m \\ X_m^T & 2\rho_{1m} I_N \otimes I_n - \rho_{1m}^2 \widetilde{R}_{mr} \end{bmatrix} \geq 0, \tag{2.50}$$

$$\begin{bmatrix} Z_m & Y_m \\ Y_m^T & 2\rho_{2m} I_N \otimes I_n - \rho_{2m}^2 \widetilde{R}_{mr} \end{bmatrix} \geq 0 \tag{2.51}$$

with

$$\widetilde{\digamma}_1 = \begin{bmatrix} I_N \otimes A - (D + H) \otimes BK & H_1^r & \cdots & H_s^r & 0 & \cdots & 0 & H^r \end{bmatrix},$$

$$\widetilde{\Xi}_m = \sqrt{\bar{d}_m} \begin{bmatrix} I_N \otimes (A - I_n) - (D + H) \otimes BK & H_1^r & \cdots & H_s^r & 0 & \cdots & 0 & H^r \end{bmatrix}$$

and $\widetilde{\Pi}_r$ is defined as: P_r is replaced with $2\rho I_N \otimes I_n - \rho^2 \widetilde{P}_r$ and other elements are the same as in Π_r in (2.42).

Proof Define the new variables as $\widetilde{P}_r \triangleq P_r^{-1}$ and $\widetilde{R}_{mr} \triangleq R_{mr}^{-1}$. Pre- and post-multiplying (2.39) with $diag\{I_{2(s[+1)N} \otimes I_n, \widetilde{P}_r, \widetilde{R}_{1r}, \ldots, \widetilde{R}_{sr}\}$ and taking into account the following inequality

$$P \geq 2\rho I - \rho^2 P^{-1}$$

holds for any positive definite matrix P and scalar $\rho > 0$, then (2.49) follows immediately. Furthermore, there exist positive scalars $\rho_{1m} > 0$ and $\rho_{2m} > 0$, $m = 1, 2, \ldots, s$ such that the satisfaction of conditions in (2.50) and (2.51) imply (2.40) and (2.41) hold. This completes the proof.

Remark 2.8 By using the switching Lyapunov functional method, sufficient conditions for the consensus are obtained in terms of LMIs in Theorem 2.2. This is reasonable at least in theory. When the number of agents in systems (2.1) is lager, the number of switching signals in Theorem 2.2 is also larger, which leads to a larger number of LMIs to be solved simultaneously. We have two methods to overcome this problem. First, we can obtain the feedback gain matrix K and the event-triggered parameter matrices Ω_i by using the event-triggered consensus control method as used in [4] without communication delays compensation, i.e. without using the predictive control scheme. Another one is use the common Lyapunov functional method to solve the feedback gain matrix K and the event-triggered parameter matrices $\Omega_i, i \in S_N$. Furthermore, we can compare the D-MB-ETPC scheme in this chapter with the ET scheme based on the same feedback gain and the event-triggered parameters to show the advantages of our proposed method.

Remark 2.9 Reference [5] has pointed out that the ETPC method in it can also counteract (compensate for) the bounded delays or information losses for single NCS. The main corresponding results are given in Theorem 4.1 (this is guaranteed by introducing the input consistent control Algorithm 1) and Corollary 4.2 (this is guaranteed by re-establishing input consistency), which are similar to the obtained results in Theorem 1 and Theorem 16 in our previous works [6, 7] for single NCS by using other ETPC schemes, respectively, where the transmission delays and data dropouts can be compensated completely. However, first, the predictive models in ref. [5] and our present chapter are different since the investigated systems are different in essence (our predictive control scheme is more complex due to the complexity of MASs); second, the analyze methods in the two papers are different. So the method proposed in work [5] cannot be directly used in our chapter to solve the consensus problem with communication delays.

2.3 Uniform Delay Case

For the better understanding of the D-MB-ETPC scheme proposed in this article, in this section we assume that the communication delays between agent i and its neighbors are all the same, that is, $\tau_{ji}(t) = \tau_i(t)$, $\forall j = 1, 2, \ldots, N$ (in practice, this can be realized by using the dwell-time method), where $\tau_i(t)$ is the network delay when agent i obtains information of itself. In other words, this section will based on the following consensus protocol

$$u_i(t) = -K\Big\{\sum_{j\in N_i} a_{ij}[x_i(t-\tau_i(t)) - x_j(t-\tau_j(t))] + b_i(x_i(t-\tau_i(t)) - x_0(t-\tau_0(t)))\Big\}$$

to investigated the D-MB-ETPC scheme.

For simplicity, we use τ_k^i to represent the transmission delay $\tau_i(t_k^i)$ for the event-triggered state $x_i(t_k^i)$ of agent i. Then, the event-triggered state $x_i(t_k^i)$ of agent i can be obtained by its neighbors simultaneously with transmission delay τ_k^i. By using the D-MB-ETPC scheme and following a similar event-triggered predictive control process analysis as in previous section, the control for NMASs (2.1) can be obtained as

$$\begin{aligned} u_i(t) &= \widehat{u}_i(t|t_k^i) \\ &= -K\Big\{\sum_{j\in N_i} a_{ij}[\delta_i(t) - A^{t-t^j_{k_j(t)}}\delta_j(t^j_{k_j(t)})] + b_i\delta_i(t)\Big\} \\ & t \in [t_k^i + \tau_k^i, t_{k+1}^i + \tau_{k+1}^i) \cap [t^j_{k_j(t)} + \tau^j_{k_j(t)}, t^j_{k_j(t)+1} + \tau^j_{k_j(t)+1}) \end{aligned} \tag{2.52}$$

where

$$k_j(t) \triangleq \arg\min_m\{t - (t_m^j + \tau_m^j))|t \geq t_m^j + \tau_m^j), m = 1, 2, \ldots\} \tag{2.53}$$

Let

$$\begin{aligned} \Theta &= [t^j_{k_j(t)} + \tau^j_{k_j(t)}, t^j_{k_j(t)+1} + \tau^j_{k_j(t)+1}) = \cup\Theta_l \\ \Theta_l &= [t^j_{k_j(t)} + l + s_l^i, t^j_{k_j(t)} + l + 1 + s_{l+1}^i) \end{aligned}$$

with $l = 0, 1, \ldots, t^j_{k_j(t)+1} - t^j_{k_j(t)} - 1$. If $l = t^j_{k_j(t)+1} - t^j_{k_j(t)} - 1$, then $s_{l+1}^i = \tau^j_{k_j(t)+1}$, otherwise, $s_l^i = \tau^j_{k_j(t)}$.

Define

$$\begin{aligned} \eta_j(t) &= t - t^j_{k_j(t)} - l \\ e_j(t - \eta_j(t)) &= \delta_j(t^j_{k_j(t)}) - \delta_j(t^j_{k_j(t)} + l) \\ & t \in \Theta_l. \end{aligned}$$

It can be easily obtained that

$$\underline{\tau}_j \leq \eta_j(t) \leq \overline{d} + 1 \triangleq \eta_M.$$

According to the definition of $k_j(t)$ in (2.53) and the event-triggered conditions in (2.3) and (2.4), we can obtain that

$$\delta_j(t^j_{k_j(t)}) = \delta_j(t - \eta_j(t)) + e_j(t - \eta_j(t)).$$

Similar to the modeling process for previous section, and by letting

$$\begin{aligned} e(t - \eta(t)) &= \left[e_1^T(t - \eta_1(t)) \; e_2^T(t - \eta_2(t)) \cdots e_N^T(t - \eta_N(t)) \right]^T, \\ \delta(t - \eta(t)) &= \left[\delta_1^T(t - \eta_1(t)) \; \delta_2^T(t - \eta_2(t)) \cdots \delta_N^T(t - \eta_N(t)) \right]^T, \end{aligned}$$

the closed-loop systems for NMASs (2.1) and (2.2) can be given as follows:

$$\begin{aligned} \delta(t+1) = {} & (I_N \otimes A - D \otimes BK - H \otimes BK)\delta(t) + (\mathcal{A} \otimes BK) \cdot \overline{A}^{\sigma(t)} \delta(t - \eta(t)) \\ & + (\mathcal{A} \otimes BK) \cdot \overline{A}^{\sigma(t)} e(t - \eta(t)) \end{aligned}$$

with $\overline{A}^{\sigma(t)}$=*diag*$\{ A^{\sigma_1(t)}, A^{\sigma_2(t)}, \ldots, A^{\sigma_N(t)} \}$, $\sigma_j(t) = t - t^j_{k_j(t)} \in \varpi \triangleq \{1, 2, \ldots, p\}$. Furthermore, the event-triggering conditions (2.3)–(2.6) together with Lemma 2.2 imply that

$$\sum_{i=1}^{N} \left[\lambda \delta_i^T(t - \eta_i(t)) \Omega_i \delta_i(t - \eta_i(t)) - e_i^T(t - \eta_i(t)) \Omega_i e_i(t - \eta_i(t)) \right] > 0.$$

The results for this case will be given in the following theorem.

Theorem 2.3 *Under the local event-triggered control laws (2.52) with the event-triggering conditions (2.3)–(2.5), for given system matrices* A, B, $\lambda > 0$, $\rho > 0$, $\rho_1 > 0$, $\rho_2 > 0$, *the group of the follower agents (2.1) achieve leader-following consensus with uniform communication delays, if there exist positive definite matrices* $\widetilde{P}_r > 0$, $Q_r > 0$, $\widetilde{R}_r > 0$, $Z > 0$, $\Omega_i > 0$, $i \in S_N$, *matrices* X *and* Y *with suitable dimensions such that the following LMIs are satisfied for all* $r = [r_1, r_2, \ldots, r_N]$, $r_i \in \varpi$,

$$\begin{bmatrix} \Pi_r + \Psi + \Psi^T + \eta_M Z & * & * \\ \digamma_1 & -\widetilde{P}_r & * \\ \Xi_1 & 0 & -\widetilde{R}_r \end{bmatrix} < 0,$$

$$\begin{bmatrix} Z & X \\ X^T & 2\rho_1 I_{Nn} - \rho_1^2 \widetilde{R}_r \end{bmatrix} \geq 0,$$
$$\begin{bmatrix} Z & Y \\ Y^T & 2\rho_2 I_{Nn} - \rho_2^2 \widetilde{R}_r \end{bmatrix} \geq 0,$$

where

$$\Pi_r = \begin{bmatrix} Q_r - 2\rho I_{Nn} + \rho^2 \widetilde{P}_r & * & * & * \\ 0 & \lambda\Omega & * & * \\ 0 & 0 & -Q_r & * \\ 0 & 0 & 0 & -\Omega \end{bmatrix},$$

$$\digamma_1 = \left[I_N \otimes A - (D+H) \otimes BK \; (\mathcal{A} \otimes BK) \cdot \overline{\mathcal{A}}^r \; 0 \; (\mathcal{A} \otimes BK) \cdot \overline{\mathcal{A}}^r \right],$$
$$\Xi_m = \sqrt{\eta_M} \left[I_N \otimes (A - I_n) - (D+H) \otimes BK \; (\mathcal{A} \otimes BK) \cdot \overline{\mathcal{A}}^r \; (\mathcal{A} \otimes BK) \cdot \overline{\mathcal{A}}^r \right],$$
$$\Psi = \left[X \; Y - X \; -Y \; 0 \right],$$

and $\Omega = diag\{\Omega_1, \Omega_2, \ldots, \Omega_N\}$.

Proof Take the switching Lyapunov functional candidate as

$$V(t, \delta(t)) = V_1(t, \delta(t)) + V_2(t, \delta(t)) + V_3(t, \delta(t))$$

with

$$V_1(t, \delta(t)) = \delta^T(t) P_r \delta(t)$$
$$V_2(t, \delta(t)) = \sum_{\theta = t - \eta_M}^{t-1} \delta^T(\theta) Q_r \delta(\theta)$$
$$V_3(t, \delta(t)) = \sum_{j=-\eta_M}^{-1} \sum_{\beta = t+j}^{t-1} \varsigma^T(\theta) R_r \varsigma(\theta)$$
$$\varsigma(\theta) = \delta(\theta + 1) - \delta(\theta),$$

where P_r, Q_r and R_r are positive definite matrices. Similar to

Remark 2.10 Considering the fact that in digital networks communication channels have only finite capabilities, that's to say, an agent can transmit only finite bits of information to its neighbors at each time step. Works [8, 9] have investigated the event-triggered distributed consensus for MASs with each pair of adjacent agents consists of a dynamic encoder-decoder pair and time-varying quantization levels, which can ensure consensus by just one bit information exchange between each pair of adjacent agents at each time step. The quantization methods under event-triggered control scheme proposed in them are worth learning for our future research.

2.4 Simulation Example

For testing and verifying the effectiveness of the proposed D-MB-ETPC scheme, we consider a group of agents with three following agents and one leader. Each agent is assumed to be a vertical taking-off and landing (VTOL) aircraft. The dynamics of the three VTOL aircraft and the leader for a typical loading and flight condition at the air speed of 135 kt [10, 11] are described by the continuous-time systems

$$\dot{x}_i(t) = \overline{A}x_i(t) + \overline{B}u_i(t) \tag{2.54}$$

$$\dot{x}_0(t) = \overline{A}x_0(t), \tag{2.55}$$

where the constant matrices $\overline{A}$ and $\overline{B}$ are given as follows

$$\overline{A} = \begin{bmatrix} -0.0366 & 0.0271 & 0.0188 & -0.4555 \\ 0.0482 & -1.01 & 0.0024 & -4.0208 \\ 0.1002 & 0.3681 & -0.707 & 1.420 \\ 0 & 0 & 1 & 0 \end{bmatrix}, \overline{B} = \begin{bmatrix} 0.4422 & 0.1761 \\ 3.5446 & -7.5922 \\ -5.52 & 4.49 \\ 0 & 0 \end{bmatrix}.$$

The state variables are: x_{i1}−horizontal velocity; x_{i2}−vertical velocity; x_{i3}−pitch rate; x_{i4}−pitch angle. The control inputs are given by u_{i1}−collective and u_{i2}−longitudinal cyclic.

Discretize the systems (2.54) and (2.55) with the sampling period $T = 0.01s$, we can obtain the systems (2.1) and (2.2) with system matrices A and B as follows:

$$A = \begin{bmatrix} 0.9996 & 0.0003 & 0.0002 & -0.0046 \\ 0.0005 & 0.9899 & 0.0000 & -0.0402 \\ 0.0010 & 0.0037 & 0.9929 & 0.0142 \\ 0 & 0 & 0.0100 & 1.0000 \end{bmatrix}, B = \begin{bmatrix} 0.0044 & 0.0018 \\ 0.0354 & -0.0759 \\ -0.0552 & 0.0449 \\ 0 & 0 \end{bmatrix}.$$

The communication topology is shown in Fig. 2.2, the corresponding adjacent matrix $\mathcal{A}$ and the degree matrix D can be easily obtained as follows:

$$\mathcal{A} = \begin{bmatrix} 0 & 1 & 1 \\ 1 & 0 & 0 \\ 0 & 1 & 0 \end{bmatrix}, D = \begin{bmatrix} 2 & 0 & 0 \\ 0 & 1 & 0 \\ 0 & 0 & 1 \end{bmatrix}.$$

Furthermore, set H as $H = diag\{1, 1, 0\}$ since agent 1 and agent 2 can access to the leader's information.

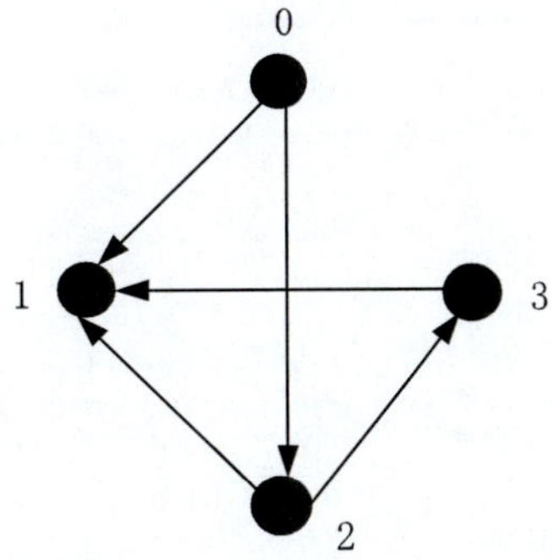

Fig. 2.2 Communication topology

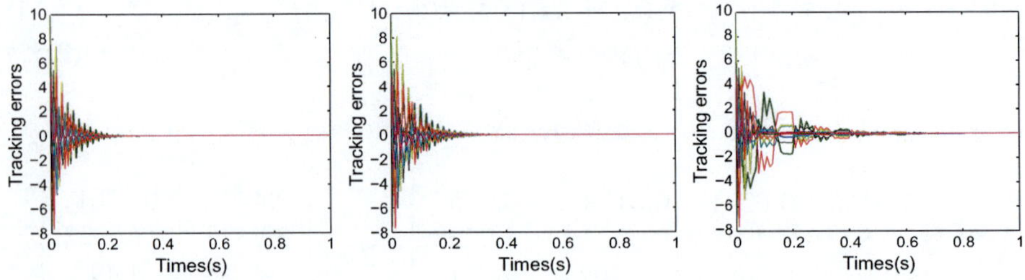

Fig. 2.3 Consensus errors of agents for Cases 1–3, respectively

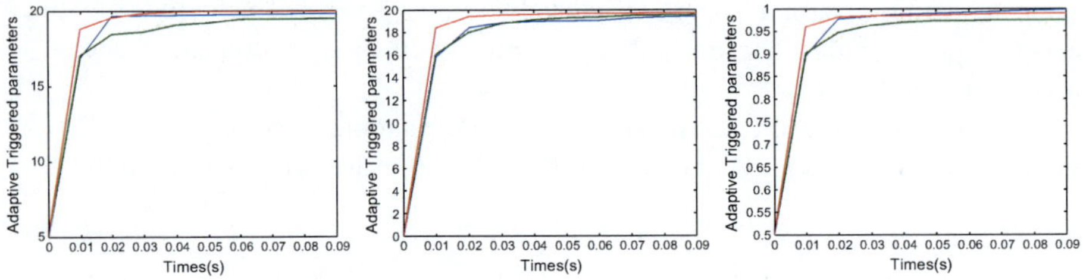

Fig. 2.4 Adaptive triggering parameters $\mu_i(t)$ for Cases 1–3, respectively

The initial conditions of each agent are set as $x_0(0) = [\,0.2\ 1\ -2\ 0.5\,]^T, x_1(0) = [\,1.2\ 2\ 1\ -0.5]^T, x_2(0) = [\,-1.8\ 7\ 2\ -5.5\,]^T, x_3(0) = [\,9.2\ -1\ 5\ -4.5\,]^T$. Assume that the communication delays for all agents are the same as $\tau = 5T$.

To show the advantages of the proposed D-MB-ETPC scheme in this chapter, we will compare our method with the traditional time-triggered predictive control scheme as has been studied in [1, 12] and the event-triggered control scheme without communication delays compensation as has been used in works [4, 13] . By solving the LMIs in Theorem 2.3 with $\mu_i(0) = 5, i = 1, 2, 3, \lambda = 20, q = 8, \rho = 0.8, \rho_1 = 0.2, \rho_2 = 0.2,$ we can obtain the feedback gain matrix K and the event-triggering parameter matrices $\Omega_i, i = 1, 2, 3$ as

$$K = \begin{bmatrix} 10.1515 & 0.7168 & -1.5310 & -7.5276 \\ 0.8106 & -0.3161 & -0.0485 & -0.0546 \end{bmatrix}, \Omega_1 = \begin{bmatrix} 0.5353 & 0.0497 & -0.0026 & -0.2820 \\ 0.0497 & 0.0084 & 0.0011 & -0.0272 \\ -0.0026 & 0.0011 & 0.0119 & 0.0189 \\ -0.2820 & -0.0272 & 0.0189 & 0.1953 \end{bmatrix}, \tag{2.56}$$

$$\Omega_2 = \begin{bmatrix} 0.5155 & -0.0015 & 0.0479 & -0.2698 \\ 0.0479 & 0.0083 & 0.0012 & -0.0261 \\ -0.0015 & 0.0012 & 0.0118 & 0.0182 \\ -0.2698 & -0.0261 & 0.0182 & 0.1879 \end{bmatrix}, \Omega_3 = \begin{bmatrix} 0.5094 & 0.0475 & -0.0012 & -0.2664 \\ 0.0475 & 0.1862 & 0.0012 & -0.0258 \\ -0.0012 & 0.0012 & 0.0117 & 0.0179 \\ -0.2664 & -0.0258 & 0.0179 & 0.1855 \end{bmatrix}. \tag{2.57}$$

Table 2.1 The sampled and released numbers for agents

	Agent 1	Agent 2	Agent 3
	Sampled numbers 100	100	100
Case 1	16	17	20
Case 2	100	100	100
Case 3	25	26	18

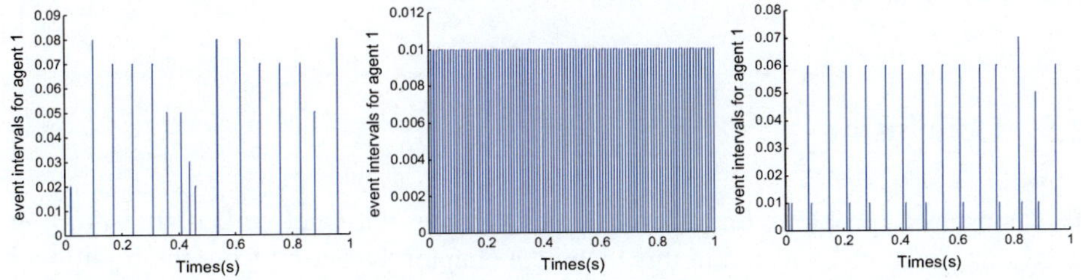

Fig. 2.5 Event times and event-intervals of agent 1 for Cases 1–3,respectively

Case 2.1 If the communication delays are compensated by using the D-MB-ETPC scheme proposed in this chapter based on the above obtained matrices in (2.56)–(2.57), the consensus errors are shown in Fig. 2.3a, from which we can see that agents can achieve consensus quickly within one second.

For fair comparison, in the following two cases we will choose the same feedback gain matrix K and event-triggering parameter matrices $\Omega_i, i = 1, 2, 3$ as obtained for Case 1.

Case 2.2 If the communication delays are compensated by using the traditional time-triggered predictive control scheme proposed in works [1, 12], $\lambda = 20$, the consensus errors are shown in Fig. 2.3b, from which we can see that the consensus performance (in terms of the convergence time and the fluctuation of the curves) is very similar to Case 1.

Case 2.3 If the communication delays are not compensated as in works [4, 13], $\mu_i(0) = 0.5$, $i = 1, 2, 3$, $\lambda = 1$, the consensus errors are drawn in Fig. 2.3c. One can see that the consensus performance is worse than the ones in Case 1 and Case 2, which shows the effectiveness of our method in compensating for the communication delays actively.

The evolutions of the adaptive triggering parameters $\mu_i(t)$ $(i = 1, 2, 3)$ for the three cases are shown in Fig. 2.4a, b, c, respectively, from which it can be seen that the triggering parameters converge to some constant values when the MASs reach consensus.

Furthermore, during the same simulation time interval $[0, 1]s$, the released numbers of the following agents for above three cases are shown in Table 2.1, the corresponding event-times and event-intervals for agent 1 (other agents have similar characteristics, so we have omitted them here) are shown in Fig. 2.5a, b, c, respectively. It is obvious that the released numbers for Case 1 and Case 3 are much less than the total sampling number 100 (Case 2). This is consistent with our theoretical results that the scheme proposed in this chapter can also significantly reduce the utilization of the communication resources. The simulation results successfully demonstrates the effectiveness of our proposed method.

2.5 Conclusion

The consensus problem for discrete-time NMASs with networked transmission delays has been investigated in this technical chapter. Different from the traditional consensus control protocols, the new D-MB-ETPC has been employed to compensate for the communication delays actively while saving the utilization of the scare communication resources for the multi-agent network. The event-triggering parameters in the event-triggered condition are adaptively adjusted according to some adaptive laws, which can overcome the drawback of the traditional event-triggering conditions with the difficulty of selecting suitable initial values. By using the time-delay system method, the virtual interval segmentation technique together with the switching Lyapunov functional analysis method, sufficient conditions in forms of linear matrix inequalities have been derived to guarantee the consensus. One simulation example is given at last to show the effectiveness of the proposed adaptive model-based event-triggered predictive control scheme.

References

1. C. Tan, G.-P. Liu, Consensus of discrete-time linear networked multi-agent systems with communication delays. IEEE Trans. Autom. Control **58**(11), 2962–2968 (2013)
2. H.-T. Zhang, Z. Cheng, G. Chen, C. Li, Model predictive flocking control for second-order multi-agent systems with input constraints. IEEE Trans. Circuits Syst I: Regular Pap. **62**(6), 1599–1606 (2015)
3. D. Yue, E. Tian, Q.-L. Han, A delay system method for designing event-triggered controllers of networked control systems. IEEE Trans. Autom. Control **58**(2), 475–481 (2012)
4. Z. Wang, M. Fei, D. Du, M. Zheng, Decentralized event-triggered average consensus for multi-agent systems in cpss with communication constraints. IEEE/CAA J. Automatica Sinica **2**(3), 248–257 (2015)
5. P. Varutti, B. Kern, T. Faulwasser, R. Findeisen, "Event-based model predictive control for networked control systems," in *Proceedings of 2009 IEEE Conference on Decision and Control* (Shanghai, 2009), pp. 567–572
6. X. Yin, D. Yue, S. Hu, Model-based event-triggered predictive control for networked systems with communication delays compensation. Int. J. Robust Nonlinear Control **25**(18), 3572–3595 (2015)

7. X. Yin, D. Yue, S. Hu, C. Peng, Y. Xue, Model-based event-triggered predictive control for networked systems with data dropout. SIAM J. Control Optim. **54**(2), 567–586 (2016)
8. H. Li, G. Chen, T. Huang, Z. Dong, W. Zhu, L. Gao, Event-triggered distributed average consensus over directed digital networks with limited communication bandwidth. IEEE Trans. Cybern. **46**(12), 3098–3110 (2016)
9. H. Li, G. Chen, T. Huang, Z. Dong, High-performance consensus control in networked systems with limited bandwidth communication and time-varying directed topologies. IEEE Trans. Neural Netw. Learn. Syst. **28**(5), 1043–1054 (2016)
10. G. Wen, Z. Duan, W. Ren, G. Chen, Distributed consensus of multi-agent systems with general linear node dynamics and intermittent communications. Int. J. Robust Nonlinear Control **24**(16), 2438–2457 (2014)
11. X. Yin, D. Yue, S. Hu, Adaptive periodic event-triggered consensus for multi-agent systems subject to input saturation. Int. J. Control **89**(4), 653–667 (2016)
12. C. Tan, G.-P. Liu, P. Shi, Consensus of networked multi-agent systems with diverse time-varying communication delays. J. Franklin Inst. **352**(7), 2934–2950 (2015)
13. C. Peng, Q.-L. Han, A novel event-triggered transmission scheme and $\mathcal{L}_2$ control co-design for sampled-data control systems. IEEE Trans. Autom. Control **58**(10), 2620–2626 (2013)

Chapter 3
Distributed Event-Triggered Tracking Control for Heterogeneous Multiagent Systems

In this chapter, we investigate the consensus problem of a set of discrete-time heterogeneous MASs with random communication delays represented by a Markov chain, where the MASs are composed of two kinds of agents differed by their dynamics. First, distributed consensus control is designed by employing the event-triggered communication technique, which can lead to a significant reduction of the information communication burden in the multi-agent network. Then, the mean square stability of the closed loop MASs is analyzed based on the Lyapunov functional method and the Kronecker product technique. Sufficient conditions are obtained to guarantee the consensus in terms of linear matrix inequalities (LMIs). Finally, a simulation example is given to illustrate the effectiveness of the developed theory.

The remainder of this chapter is shown as follows. In Sect. 3.1, we present the consensus problem to be investigated in this chapter and the discrete-time event-triggered control scheme. The main result of this chapter is given in Sect. 3.2. Simulation results are presented to illustrate the effectiveness of the theoretical results in Sect. 3.3. Conclusions are drawn in Sect. 3.4.

3.1 Problem Formulation

Suppose that the heterogeneous MASs considered in this chapter consist of n agents indexed by $1, 2, \ldots n$. The first l $(l < n)$ agents are assumed to be second-order agents and the last $(n - l)$ agents are assumed to be first-order agents. The dynamics of the second order agents can be described as

$$\begin{aligned} x_i(k+1) &= x_i(k) + v_i(k), \\ v_i(k+1) &= v_i(k) + u_i(k), i = 1, 2, \ldots, l, \end{aligned} \tag{3.1}$$

D. Yue et al., *Distributed Cooperative Control and Communication for Multi-agent Systems*, https://doi.org/10.1007/978-981-33-6718-0_3

and the first-order agents can be described as

$$x_i(k+1) = x_i(k) + u_i(k), i = l+1, l+2, \ldots, n, \tag{3.2}$$

where $x_i(k) \in R^p$, $v_j(k) \in R^p$, $u_i(k) \in R^p$, $i \in S_n$, $j \in S_l$ with p being the preassigned dimension denote the state vector, the velocity vector and the control vector of the ith agent, respectively. The initial conditions are $x_i(0) = x_{i0}$, $v_i(0) = v_{i0}$. Let $x^T(0) = \left[x_{10}^T, x_{20}^T, \ldots, x_{n0}^T\right]$ and $v^T(0) = \left[v_{10}^T, v_{20}^T, \ldots, v_{l0}^T\right]$.

Assumption 3.1 The weighted directed graph for the communication network contains a directed spanning tree.

Remark 3.1 Assumption 3.1 means that, in the set of agents, there exist at least one agent can be able to impact on other agents, if not, there may exist one subgroup of agents, which can not receive any information from others. Therefore, the consensus of such a group of agents may never be achieved. One can refer to [1] for details.

3.2 Event-Triggered Tracking Control Design

In this subsection, the consensus control algorithm will be designed based on model (3.1) and model (3.2) with considering the effect of communication delays and the event-triggered communication scheme.

In many cases, delays frequently occurred in practical systems [2, 5, 7] . For the control design of MASs, two type of consensus protocols with communication delays have been considered by now. One is that the delay only affects the state received from neighboring agents. Another is that the delay affects both the state received from itself and the state received from its neighbors [3]. In this chapter, the consensus problem is dealt with in the presence of uniform and random delays only affect the state received from neighboring agents. All the communication channels among the agents are assumed to subject to the same delay at one instant. Here, we say the consensus is achieved if and only if all the states in (3.1) and (3.2) converge to b_0, where b_0 is called the static leader, denotes the prescribed rendezvous point. For the help of giving the event-triggered control scheme next, we first give the following linear consensus control law for the heterogeneous MASs (3.1) and (3.2) as:

$$u_i(k) = -\gamma v_i(k) - K \sum_{j \in N_i} \left[a_{(i,j)}(x_i(k) - x_j(k-d_k))\right] + \beta_i(b_0 - x_i(k)), i \in S_l, \tag{3.3}$$

$$u_i(k) = -K \sum_{j \in N_i} a_{(i,j)}(x_i(k) - x_j(k-d_k)) + \beta_i(b_0 - x_i(k)), i = l+1, l+2, \ldots, n, \tag{3.4}$$

where $\gamma > 0$ is the gain and K is the feedback matrix to be determined. d_k is the communication delay at instant k. For generally, $\beta_i \geq 0$ is the coupling weight between the agent i and the static leader b_0 with the characteristic that $\beta_i > 0$ if agent i is connected to the leader, otherwise, $\beta_i = 0$. But in this chapter, since the event-triggered condition proposed later in (3.7) for each agent depends not only on its local state information but also depends on the value of b_0, the value of b_0 should be settled into every agent previously. Under this assumption, the calculation and communication cost will not increase by introducing the static point of b_0.

For the use of the following sections, we define $H = diag\{H_1, H_2\}$ with $H_1 \triangleq diag\{\beta_1, \beta_2, \ldots, \beta_l\}$ and $H_2 \triangleq diag\{\beta_{l+1}, \beta_{l+2}, \ldots, \beta_n\}$.

The communication delay d_k in (3.3) and (3.4) is random and governed by a homogeneous Markov chain. The definition of homogeneous Markov chain is given as follows.

Definition 3.1 ([1]) The Markov chain X with space S is called homogeneous if

$$\mathbb{P}(X_{m+1} = j|X_m = i) = \mathbb{P}(X_1 = j|X_0 = i)$$

for all m, i, j. The transition matrix $\Pi = (\pi_{ij})$ is the $|S| \times |S|$ matrix of transition probabilities

$$\pi_{ij} = \mathbb{P}(X_{m+1} = j|X_m = i),$$

where $\mathbb{P}$ is the probability operator and $|S|$ is the cardinality of state space S. The transition probability matrix Π satisfies

$$\sum_{j=1}^{|S|} \pi_{ij} = 1, i = 1, 2, \ldots, |S|.$$

Assumption 3.2 The time delay d_k takes values in $\Gamma = \{\tau_1, \tau_2, \ldots, \tau_q\}$, where τ_k $(k = 1, 2, \ldots, q)$ are integers and satisfy $0 \le \tau_1 \le \tau_2 \le \cdots \le \tau_q$.

Remark 3.2 The dynamics of the systems may change smoothly during a finite time interval in many real systems. Therefore, the time-triggered communication scheme as in (3.3)–(3.4) may produce many useless messages communicated through the network, which then leads to a conservative usage of the communication resources and bandwidth. In order to reduce the communication burden and save the communication energy of the multi-agent network, in the following section, the event-triggered communication scheme is first introduced to realize the consensus controller (3.3)–(3.4), where the events are triggered for each agent i when some trigger function is positive.

The realizing process for the algorithm with the consensus controller (3.3)–(3.4) under an event-triggered communication scheme can be illustrated as follows: Each agent computes its trigger function with the current obtained state signal $x_i(k)$ and the last released state signal $\widetilde{x}_i(k)$ (this will be defined later). If the trigger condition is fulfilled for one agent, the agent will broadcast its actual state signal to its neighbors. Under this scenario, the event-triggered consensus control is more preferable than that proposed in (3.3)–(3.4). The time when the agent sends its state signal out to its neighbors is called the release time (or the event time).

Now, we redefine the above control formulations (3.3) and (3.4) to take into account the event-triggered communication strategy. It is more practical to consider the distributed event-triggered communication scheme. In order to model the event-triggers for agents, here we first denote the state release times for the ith agent and its neighbors by s_m^i and s_m^j ($m = 0, 1, 2, ...; i \in S_n; j \in N_i$), respectively. Then the broadcasted states for each agent i can be described by the following piece-wise constant functions:

$$\widetilde{x}_i(k) = x_i(s_m^i), k \in [s_m^i, s_{m+1}^i), i \in S_n, \tag{3.5}$$

Define the trigger functions as $f_i(\cdot, \cdot, \cdot), i \in S_n$ which depend on local information of agent i only. An event for agent i is triggered as soon as the following relationship

$$f_i(x_i(k), \widetilde{x}_i(k), b_0) < 0 \tag{3.6}$$

is violated. In this chapter, the detailed trigger functions are defined as

$$f_i(x_i(k), \widetilde{x}_i(k), b_0) = e_i^T(k)\Omega_i e_i(k) - \sigma_i \delta_i^T(k)\Omega_i \delta_i(k), i \in S_n, \tag{3.7}$$

where

$$\begin{aligned}
\delta_i(k) &= x_i(k) - b_0 \\
\widetilde{\delta}_i(k) &= \widetilde{x}_i(k) - b_0 \\
e_i(k) &= \widetilde{\delta}_i(k) - \delta_i(k), i \in S_n,
\end{aligned} \tag{3.8}$$

and $\sigma_i, i \in S_n$ are given positive scalars which satisfy $0 < \sigma_i < 1$ and $\Omega_i, i \in S_n$ are positive definite trigger-matrices to be designed later to guarantee the consensus. It should be noted that the trigger functions take values in R.

Then, the consensus algorithm with the controller (3.3) and (3.4) can be implemented replacing the period signals by these piecewise constant functions (3.5) under the event-triggered communication scheme, called the event-triggered consensus algorithm or the event-triggered control, which can be described as

$$\begin{aligned}
u_i(k) &= -\gamma v_i(k) - K \sum_{j \in N_i} \left[a_{(i,j)}(x_i(s_m^i) - x_j(s_{m_j(k-d_k)}^j))\right] + \beta_i(b_0 - x_i(s_m^i)), \\
&= -\gamma v_i(k) - K \sum_{j \in N_i} \left[a_{(i,j)}(\widetilde{x}_i(k) - \widetilde{x}_j(k - d_k))\right] + \beta_i(b_0 - \widetilde{x}_i(k)), i \in S_l
\end{aligned} \tag{3.9}$$

$$\begin{aligned}
u_i(k) &= -K \sum_{j \in N_i} a_{(i,j)}(x_i(s_m^i) - x_j(s_{m_j(k-d_k)}^j)) + \beta_i(b_0 - x_i(s_m^i)) \\
&= -K \sum_{j \in N_i} a_{(i,j)}(\widetilde{x}_i(k) - \widetilde{x}_j(k - d_k)) + \beta_i(b_0 - \widetilde{x}_i(k)), i = l+1, \ldots, n
\end{aligned} \tag{3.10}$$

where

$$m_j(k) \triangleq arg \min_{m \in N: s_m^j \le k} \left\{ k - s_m^j \right\}, \tag{3.11}$$

and $k \in [s_m^i, s_{m+1}^i)$, $N = \{1, 2, 3, \ldots\}$.

Remark 3.3 Combining (3.8) and the definition of $m_j(k)$ in (3.11), it can be easily obtained that

$$\delta_j(s_{m_j(k)}^j) = \delta_j(k) + e_j(k),$$

furthermore, by taking into account the communication delays, we have

$$\delta_j(s_{m_j(k-d_k)}^j) = \delta_j(k - d_k) + e_j(k - d_k). \tag{3.12}$$

Equation (3.12) will be used later.

Combining (3.1)–(3.2) and (3.9)–(3.10) with $\delta_i(k) = x_i(k) - b_0$, we can obtain the dynamics of $\delta_i(k)$ and $v_i(k)$ as follows:

$$\delta_i(k+1) = \delta_i(k) + v_i(k), \tag{3.13}$$

$$v_i(k+1) = v_i(k) - \gamma v_i(k) - K \sum_{j \in N_i} \left[a_{(i,j)}(\widetilde{\delta}_i(k) - \widetilde{\delta}_j(k - d_k)) \right] \tag{3.14}$$

$$-\beta_i \widetilde{\delta}_i(k), i \in S_l \tag{3.15}$$

$$\delta_i(k+1) = \delta_i(k) - K \sum_{j \in N_i} a_{(i,j)}(\widetilde{\delta}_i(k) - \widetilde{\delta}_j(k - d_k)) - \beta_i \widetilde{\delta}_i(k), i = l+1, l+2, \ldots, n. \tag{3.16}$$

Definition 3.2 ([1]) For the discrete-time heterogeneous MASs (3.1) and (3.2), protocols u_i in (3.9) and (3.10) are said to mean square asymptotically solve the consensus problem if and only if for any initial values we have

$$\lim_{k \to \infty} E\left\{ \|x_i(k) - b_0\|^2 \right\} = 0, i \in S_n,$$

$$\lim_{k \to \infty} E\left\{ \|v_i(k)\|^2 \right\} = 0, i \in S_l.$$

The following Lemmas are helpful to derive the main result of this chapter.

Lemma 3.1 ([6]) *The Kronecker product $\otimes$ has the following properties: for matrices A, B, C and D of appropriate dimensions,*
(1) $(A + B) \otimes C = A \otimes C + B \otimes C$;
(2) $(A \otimes B)(C \otimes D) = AC \otimes BD$.

Lemma 3.2 *For real matrices X and Y and $S = S^T > 0$ with appropriate dimensions, the following inequality holds:*

$$X^T Y + Y^T X \leq X^T S X + Y^T S^{-1} Y.$$

Proof Since $S = S^T > 0$, for real matrices X and Y with appropriate dimensions, we have

$$(SX - Y)^T S^{-1}(SX - Y) \geq 0$$

and this lemma can be easily obtained by the above inequality.

3.3 Consensus Analysis

The objective of this chapter is to find the conditions for the discrete-time heterogeneous MASs (3.1) and (3.2) to reach consensus under the event-triggered communication scheme described above. Before we obtain the main results of this chapter, some algebraic manipulation is needed. Firstly, we define the following augmented variables as

$$\begin{aligned}
W(k) &= \left[v_1^T(k)\ v_2^T(k) \cdots v_l^T(k) \right]^T \in R^{lp}, \\
\Delta_1(k) &= \left[\delta_1^T(k)\ \delta_2^T(k) \cdots \delta_l^T(k) \right]^T \in R^{lp}, \\
\Delta_2(k) &= \left[\delta_{l+1}^T(k)\ \delta_{l+2}^T(k) \cdots \delta_n^T(k) \right]^T \in R^{(n-l)p}, \\
E_1(k) &= \left[e_1^T(k)\ e_2^T(k) \cdots e_l^T(k) \right]^T \in R^{lp}, \\
E_2(k) &= \left[e_{l+1}^T(k)\ e_{l+2}^T(k) \cdots e_n^T(k) \right]^T \in R^{(n-l)p}, \\
E(k) &= \left[E_1^T(k)\ E_2^T(k) \right]^T \in R^{np}.
\end{aligned}$$

and combining (3.8) and (3.13)–(3.16), we can obtain

$$\Delta_1(k+1) = \Delta_1(k) + W(k), \tag{3.17}$$

$$W(k+1) = (1-\gamma)W(k) + \begin{bmatrix} -K\sum_{j\in N_1} a_{(1,j)}(\widetilde{\delta}_1(k) - \widetilde{\delta}_j(k-d_k)) - \beta_1\widetilde{\delta}_1(k) \\ \vdots \\ -K\sum_{j\in N_l} a_{(l,j)}(\widetilde{\delta}_l(k) - \widetilde{\delta}_j(k-d_k)) - \beta_l\widetilde{\delta}_l(k) \end{bmatrix}$$

$$
\begin{aligned}
&= (1-\gamma)W(k) + \\
&\begin{bmatrix} -K\sum_{j\in N_1} a_{(1,j)}(\delta_1(k)+e_1(k)-\delta_j(k-d_k)-e_j(k-d_k)) - \beta_1(\delta_1(k)+e_1(k)) \\ \vdots \\ -K\sum_{j\in N_l} a_{(l,j)}(\delta_l(k)+e_l(k)-\delta_j(k-d_k)-e_j(k-d_k)) - \beta_l(\delta_l(k)+e_l(k)) \end{bmatrix} \\
&= (1-\gamma)W(k) - (D_1\otimes K)\Delta_1(k) + (A_1\otimes K)\Delta_1(k-d_k) \\
&\quad + (A_2\otimes K)\Delta_2(k-d_k) - (H_1\otimes I_p)\Delta_1(k) \\
&\quad - (D_1\otimes K)E_1(k) + (A_1\otimes K)E_1(k-d_k) + (A_2\otimes K)E_2(k-d_k) - \left(H_1\otimes I_p\right)E_1(k),
\end{aligned} \tag{3.18}
$$

$$
\Delta_2(k+1) = \Delta_2(k) + \begin{bmatrix} -K\sum_{j\in N_{l+1}}\left[a_{(l+1,j)}(\widetilde{\delta}_{l+1}(k)-\widetilde{\delta}_j(k-d_k))\right] - \beta_{l+1}\widetilde{\delta}_{l+1}(k) \\ \vdots \\ -K\sum_{j\in N_n}\left[a_{(n,j)}(\widetilde{\delta}_n(k)-\widetilde{\delta}_j(k-d_k))\right] - \beta_n\widetilde{\delta}_n(k) \end{bmatrix}
$$

$$
\begin{aligned}
&= \Delta_2(k) + \\
&\begin{bmatrix} -K\sum_{j\in N_{l+1}} a_{(l+1,j)}(\delta_{l+1}(k)+e_{l+1}(k)-\delta_j(k-d_k)-e_j(k-d_k)) - \beta_{l+1}(\delta_{l+1}(k)+e_{l+1}(k)) \\ \vdots \\ -K\sum_{j\in Nn} a_{(n,j)}(\delta_n(k)+e_n(k)-\delta_j(k-d_k)-e_j(k-d_k)) - \beta_n(\delta_n(k)+e_n(k)) \end{bmatrix} \\
&= \Delta_2(k) - (D_2\otimes K)\Delta_2(k) + (A_3\otimes K)\Delta_1(k-d_k) \\
&\quad + (A_4\otimes K)\Delta_2(k-d_k) - \left(H_2\otimes I_p\right)\Delta_2(k) \\
&\quad - (D_2\otimes K)E_2(k) + (A_3\otimes K)E_1(k-d_k) \\
&\quad + (A_4\otimes K)E_2(k-d_k) - \left(H_2\otimes I_p\right)E_2(k),
\end{aligned} \tag{3.19}
$$

where the notation $\otimes$ stands for the Kronecker product and $A_1.A_2$, A_3, A_4, D_1 and D_2 in (3.18) and (3.19) satisfy

$$
L = \begin{bmatrix} D_1 - A_1 & -A_2 \\ -A_3 & D_2 - A_4 \end{bmatrix},
$$

where

$$
\begin{aligned}
D_1 &= diag\left\{\sum_{j\in N_1} a_{(1,j)}, \sum_{j\in N_2} a_{(2,j)}, \ldots, \sum_{j\in N_l} a_{(l,j)}\right\}, \\
D_2 &= diag\left\{\sum_{j\in N_{l+1}} a_{(l+1,j)}, \sum_{j\in N_{l+2}} a_{(l+2,j)}, \ldots, \sum_{j\in N_n} a_{(n,j)}\right\}.
\end{aligned}
$$

In order to write (3.17)–(3.19) in a matrix form, define $Z(k) = \left[\, W^T(k)\ \Delta_1^T(k)\ \Delta_2^T(k)\,\right]^T \in R^{(l+n)p}$. The network dynamics (3.13)–(3.16) can be simplified in the following form

$$
\begin{aligned}
Z(k+1) &= \begin{bmatrix} (1-\gamma)I_{lp} - D_1 \otimes K - H_1 \otimes I_p & & 0 \\ I_{lp} & I_{lp} & 0 \\ 0 & 0 & I_{(n-l)p} - D_2 \otimes K - H_2 \otimes I_p \end{bmatrix} Z(k) \\
&+ \begin{bmatrix} 0 & A_1 \otimes K & A_2 \otimes K \\ 0 & 0 & 0 \\ 0 & A_3 \otimes K & A_4 \otimes K \end{bmatrix} Z(k-d_k) + \begin{bmatrix} A_1 \otimes K & A_2 \otimes K \\ 0 & 0 \\ A_3 \otimes K & A_4 \otimes K \end{bmatrix} E(k-d_k) \\
&+ \begin{bmatrix} -D_1 \otimes K - H_1 \otimes I_p & 0 \\ 0 & 0 \\ 0 & -D_2 \otimes K - H_2 \otimes I_p \end{bmatrix} E(k) \\
&\triangleq B_1 Z(k) + C_1 E(k) + N K_1 Z(k-d_k) + M K_2 E(k-d_k)
\end{aligned} \tag{3.20}
$$

where

$$
K_1 = \begin{bmatrix} I_l \otimes K & 0 & 0 \\ 0 & I_l \otimes K & 0 \\ 0 & 0 & I_{n-l} \otimes K \end{bmatrix}, K_2 = \begin{bmatrix} I_l \otimes K & 0 \\ 0 & I_{n-l} \otimes K \end{bmatrix}, \tag{3.21}
$$

$$
B_1 = B + F_1 K_1, C_1 = C + F_2 K_2,
$$

$$
B = \begin{bmatrix} (1-\gamma)I_{lp} - H_1 \otimes I_p & & 0 \\ I_{lp} & I_{lp} & 0 \\ 0 & 0 & I_{(n-l)p} - H_2 \otimes I_p \end{bmatrix}, F_1 = \begin{bmatrix} 0 & -D_1 \otimes I_p & 0 \\ 0 & 0 & 0 \\ 0 & 0 & -D_2 \otimes I_p \end{bmatrix},
$$

$$
C = \begin{bmatrix} -H_1 \otimes I_p & 0 \\ 0 & 0 \\ 0 & -H_2 \otimes I_p \end{bmatrix}, F_2 = \begin{bmatrix} -D_1 \otimes I_p & 0 \\ 0 & 0 \\ 0 & -D_2 \otimes I_p \end{bmatrix},
$$

$$
N = \begin{bmatrix} 0 & A_1 \otimes I_p & A_2 \otimes I_p \\ 0 & 0 & 0 \\ 0 & A_3 \otimes I_p & A_4 \otimes I_p \end{bmatrix}, M = \begin{bmatrix} A_1 \otimes I_p & A_2 \otimes I_p \\ 0 & 0 \\ A_3 \otimes I_p & A_4 \otimes I_p \end{bmatrix}.
$$

Based on the preceding preparations, sufficient conditions will be derived to guarantee the consensus of MASs (3.1) and (3.2) under event-triggered control scheme (3.9) and (3.10) with random communication delays. Taking into account (3.20)–(3.21) and by using the Lyapunov functional method and the Kronecker product technique, an LMI-based method of designing the trigger parameter matrices in (3.7) and the feedback matrix K in (3.9) and (3.10) will be given in the following Theorem.

Theorem 3.1 *Suppose that the communication topology G is directed and Assumption 3.1 and Assumption 3.2 hold, then the event-triggered control protocols (3.9) and (3.10) with given trigger parameters $\sigma_i (i \in S_n)$ in (3.7) can solve the consensus problem for the discrete-time heterogeneous MASs (3.1) and (3.2) with random delays governed by a Markov chain under Definition 3.1 if there exist symmetrical matrices $P > 0$ where $P \in R^{(l+n)p \times (l+n)p}$, $Q_j > 0$, $R_j > 0$, $Z_j > 0$, $j = 1, 2, \ldots, q$, matrices $\Omega_i > 0$, $i = 1, 2, \ldots, n$ and M_j, $j = 1, 2, \ldots, q$ with suitable dimensions such that the following inequalities are satisfied:*

$$\begin{bmatrix} \Phi_{11}(r) & * \\ \Phi_{21} & \Phi_{22} \end{bmatrix} < 0 \tag{3.22}$$

for $\forall\, r = 1, 2, \ldots, q$, *where*

$$\Phi_{11}(r) = \Theta(r) + \Psi + \Psi^T \tag{3.23}$$

$$\Theta(r) =$$

$$\begin{bmatrix} \Theta_{11} & * & * & * & * & * & * & * & * & * \\ \Theta^r_{21} & \Theta^r_{22} & * & * & * & * & * & * & * & * \\ \Theta^r_{31} & 0 & \Theta^r_{33} & * & * & * & * & * & * & * \\ \vdots & \vdots & \vdots & \vdots & * & * & * & * & * & * \\ \Theta^r_{(q+1)1} & 0 & 0 & \cdots & \Theta_{(q+1)(q+1)} & * & * & * & * & * \\ \Theta^r_{(q+2)1} & \Theta^r_{(q+2)2} & \Theta^r_{(q+2)3} & \cdots & \Theta^r_{(q+2)(q+1)} & \Theta_{(q+2)(q+2)} & * & * & * & * \\ \Theta^r_{(q+3)1} & \Theta^r_{(q+3)2} & 0 & \cdots & 0 & \Theta^r_{(q+3)(q+2)} & \Theta^r_{(q+3)(q+3)} & * & * & * \\ \Theta^r_{(q+4)1} & 0 & \Theta^r_{(q+4)3} & \cdots & 0 & \Theta^r_{(q+4)(q+2)} & 0 & \Theta^r_{(q+4)(q+4)} & * & * \\ \vdots & \vdots & \vdots & \cdots & \vdots & \vdots & \vdots & \vdots & \vdots & * \\ \Theta^r_{(2q+2)1} & 0 & 0 & \cdots & \Theta^r_{(2q+2)(q+1)} & \Theta^r_{(2q+2)(q+2)} & 0 & 0 & \cdots & \Theta^r_{(2q+2)(2q+2)} \end{bmatrix},$$

with

$$\begin{aligned}
\Theta_{11} &= (B + F_1K_1)^T\, P\,(B + F_1K_1) - P + \overline{\Omega} \\
&\quad + \sum_{j=1}^{q} Q_j + (B - I + F_1K_1)^T \overline{Z}(B - I + F_1K_1), \\
\Theta^r_{(i+1)1} &= \pi_{ri} K_1^T N^T P\,(B + F_1K_1) + \pi_{ri} K_1^T N^T \overline{Z}(B - I + F_1K_1), i = 1, 2, \ldots, q, \\
\Theta^r_{(i+1)(i+1)} &= \pi_{ri} K_1^T N^T PNK_1 + \pi_{ri} K_1^T N^T \overline{Z} N K_1 - Q_i, i = 1, 2, \ldots, q, \\
\Theta^r_{(q+2)1} &= (C + F_2K_2)^T\, P\,(B + F_1K_1) + (C + F_2K_2)^T \overline{Z}(B - I + F_1K_1), \\
\Theta^r_{(q+2)(j+1)} &= \pi_{rj}(C + F_2K_2)^T PNK_1 + \pi_{rj}(C + F_2K_2)^T \overline{Z} N K_1,\ j = 1, 2, \ldots, q, \\
\Theta_{(q+2)(q+2)} &= (C + F_2K_2)^T P(C + F_2K_2) - \Omega + \sum_{j=1}^{q} R_j + (C + F_2K_2)^T \overline{Z}(C + F_2K_2), \\
\Theta^r_{(q+2+i)1} &= \pi_{ri} K_2^T M^T P\,(B + F_1K_1) + \pi_{ri} K_2^T M^T \overline{Z}(B - I + F_1K_1), i = 1, 2, \ldots, q, \\
\Theta_{(q+2+i)(i+1)} &= \pi_{ri} K_2^T M^T PNK_1 + \pi_{ri} K_2^T M^T \overline{Z} N K_1, i = 1, 2, \ldots, q, \\
\Theta_{(q+2+i)(q+2)} &= \pi_{ri} K_2^T M^T P(C + F_2K_2) + \pi_{ri} K_2^T M^T \overline{Z}(C + F_2K_2), i = 1, 2, \ldots, q, \\
\Theta_{(q+2+i)(q+2+i)} &= \pi_{ri} K_2^T M^T PMK_2 + \pi_{ri} K_2^T M^T \overline{Z} M K_2 - R_i, i = 1, 2, \ldots, q,
\end{aligned}$$

$$\Psi = \left[\sum_{j=1}^{q} M_q \; -M_1 \; -M_2 \cdots -M_q \; 0 \; 0 \cdots 0 \right],$$

$$\Phi_{21} = \left[\sqrt{\tau_1} M_1 \; \sqrt{\tau_2} M_2 \cdots \sqrt{\tau_q} M_q \right]^T, \tag{3.24}$$

$$\Phi_{22} = -diag \left\{ Z_1, Z_2, \ldots, Z_q \right\}, \tag{3.25}$$

$$\overline{Z} = \sum_{j=1}^{q} \tau_j Z_j, \tag{3.26}$$

and

$$\overline{\Omega} = diag \left\{ 0, \ldots, 0, \sigma_1 \Omega_1, \ldots, \sigma_l \Omega_l, \sigma_{l+1} \Omega_{l+1}, \ldots, \sigma_n \Omega_n \right\},$$
$$\Omega = diag \left\{ \Omega_1, \Omega_2, \ldots, \Omega_l, \Omega_{l+1}, \ldots, \Omega_n \right\}.$$

Proof Consider the following Lyapunov functional candidate

$$\begin{aligned}
V(k) &= V_1(k) + V_2(k) + V_3(k) + V_4(k), \qquad (3.27)\\
V_1(k) &= Z^T(k) P Z(k),\\
V_2(k) &= \sum_{j=1}^{q} \sum_{i=k-\tau_j}^{k-1} Z^T(i) Q_j Z(i),\\
V_3(k) &= \sum_{j=1}^{q} \sum_{i=k-\tau_j}^{k-1} E^T(i) R_j E(i),\\
V_4(k) &= \sum_{j=1}^{q} \sum_{i=-\tau_i}^{-1} \sum_{m=k+i}^{k-1} \eta^T(m) Z_j \eta(m),\\
\eta(m) &= Z(m+1) - Z(m),
\end{aligned}$$

where P, Q_j, R_j and Z_j, $j = 1, 2, \ldots, q$ are positive definite matrices to be determined. Define [1]

$$d_{k-1} = \tau_r, d_k = \tau_s, r, s \in \{1, 2, \ldots, q\},$$

then the transition probability from d_{k-1} to d_k is $\mathbb{P}(d_{k-1} = \tau_r | d_k = \tau_s) = \pi_{rs}$. For any given k, calculating the difference of $v_i(k)$ $(i = 1, 2, 3, 4)$ in (3.27) along the trajectories of (3.20) and taking the mathematical expectation, we have

$$\begin{aligned}
&E\{\Delta V_1(k)\} = E\{V_1(k+1) - V_1(k)\}\\
&= E\Big\{[B_1 Z(k) + C_1 E(k) + N K_1 Z(k-d_k) + M K_2 E(k-d_k)]^T P\\
&[B_1 Z(k) + C_1 E(k) + N K_1 Z(k-d_k) + M K_2 E(k-d_k)] - Z^T(k) P Z(k)\Big\}\\
&= 2 \sum_{s=1}^{q} \Big[\pi_{rs} Z^T(k-\tau_s) K_1^T N^T P B_1 Z(k) + \pi_{rs} Z^T(k-\tau_s) K_1^T N^T P C_1 E(k)\\
&+ \pi_{rs} E^T(k-\tau_s) K_2^T M^T P B_1 Z(k) + \pi_{rs} E^T(k-\tau_s) K_2^T M^T P C_1 E(k)\Big]
\end{aligned}$$

$$
\begin{aligned}
&+\sum_{s=1}^{q}\Big[\pi_{rs}Z^T(k-\tau_s)K_1^TN^TPNK_1Z(k-\tau_s)+\pi_{rs}Z^T(k-\tau_s)K_1^TN^TPMK_2E(k-\tau_s)\\
&+\pi_{rs}E^T(k-\tau_s)K_2^TM^TPNK_1Z(k-\tau_s)+\pi_{rs}E^T(k-\tau_s)K_2^TM^TPMK_2E(k-\tau_s)\Big]\\
&+Z^T(k)B_1^TPB_1Z(k)+Z^T(k)B_1^TPC_1E(k)+E^T(k)C_1^TPB_1Z(k)\\
&+E^T(k)C_1^TPC_1E(k)-Z^T(k)PZ(k),
\end{aligned}
\tag{3.28}
$$

$$
E\{\Delta V_2(k)\}=\sum_{j=1}^{q}\Big[Z^T(k)Q_jZ(k)-Z^T(k-\tau_j)Q_jZ(k-\tau_j)\Big],
\tag{3.29}
$$

$$
E\{\Delta V_3(k)\}=\sum_{j=1}^{q}\Big[E^T(k)R_jE(k)-E^T(k-\tau_j)R_jE(k-\tau_j)\Big],
\tag{3.30}
$$

$$
E\{\Delta V_4(k)\}=\sum_{j=1}^{q}\eta^T(k)\tau_jZ_j\eta(k)-\sum_{j=1}^{q}\sum_{i=k-\tau_j}^{k-1}\eta^T(i)Z_j\eta(i).
\tag{3.31}
$$

For further analysis, we define the following augmented vector as

$$
\begin{aligned}
\zeta^T(k)=\big[\,&Z^T(k)\ Z^T(k-\tau_1)\ Z^T(k-\tau_2)\cdots Z^T(k-\tau_q)\\
&E(k)\ E^T(k-\tau_1)\ E^T(k-\tau_2)\cdots E^T(k-\tau_q)\big].
\end{aligned}
$$

For any matrices

$$
M_j=\big[\,M_{oj}^T\ M_{1j}^T\ M_{2j}^T\cdots M_{qj}^T\ M_{(q+1)j}^T\ M_{(q+2)j}^T\ M_{(q+3)j}^T\cdots M_{(2q+1)j}^T\,\big]^T
$$

with appropriate dimensions, it can be easily concluded that

$$
2\sum_{j=1}^{q}\zeta^T(k)M_j\left[Z(k)-Z(k-\tau_j)-\sum_{l=k-\tau_j}^{k-1}\eta(l)\right]=0
$$

and

$$
\sum_{j=1}^{q}\sum_{i=k-\tau_j}^{k-1}\left[\zeta^T(k)M_j+\eta^T(k)Z_j\right]Z_j^{-1}\left[M_j^T\zeta(k)+Z_j\eta(k)\right]\geq 0.
$$

Combining (3.6)–(3.7) and (3.28)–(3.31) and through some simple calculation, it can be obtained that

$$
\begin{aligned}
&E\{\Delta V(k)\}=E\{\Delta V_1(k)\}+E\{\Delta V_2(k)\}+E\{\Delta V_3(k)\}+E\{\Delta V_4(k)\}\\
&\leq E\{\Delta V_1(k)\}+E\{\Delta V_2(k)\}+E\{\Delta V_3(k)\}+E\{\Delta V_4(k)\}\\
&+2\sum_{j=1}^{q}\zeta^T(k)M_j\left[Z(k)-Z(k-\tau_j)-\sum_{l=k-\tau_j}^{k-1}\eta(l)\right]
\end{aligned}
$$

$$
\begin{aligned}
&+\sum_{j=1}^{q}\sum_{i=k-\tau_j}^{k-1}\left[\zeta^T(k)M_j+\eta^T(k)Z_j\right]Z_j^{-1}\left[M_j^T\zeta(k)+Z_j\eta(k)\right]\\
&+\max_{i\in S_n}\left\{\sigma_i\delta_i^T(k)\Omega_i\delta_i(k)-e_i^T(k)\Omega_ie_i(k)\right\}\\
&\le E\{\Delta V_1(k)\}+E\{\Delta V_2(k)\}+E\{\Delta V_3(k)\}+E\{\Delta V_4(k)\}\\
&+2\sum_{j=1}^{q}\zeta^T(k)M_j\left[Z(k)-Z(k-\tau_j)-\sum_{l=k-\tau_j}^{k-1}\eta(l)\right]\\
&+\sum_{j=1}^{q}\sum_{i=k-\tau_j}^{k-1}\left[\zeta^T(k)M_j+\eta^T(k)Z_j\right]Z_j^{-1}\left[M_j^T\zeta(k)+Z_j\eta(k)\right]\\
&+\sum_{i=1}^{n}\sigma_i\delta_i^T(k)\Omega_i\delta_i(k)-\sum_{i=1}^{n}e_i^T(k)\Omega_ie_i(k)\\
&=2\sum_{s=1}^{q}\Big[\pi_{rs}Z^T(k-\tau_s)K_1^TN^TPB_1Z(k)+\pi_{rs}Z^T(k-\tau_s)K_1^TN^TPC_1E(k)\\
&+\pi_{rs}E^T(k-\tau_s)K_2^TM^TPB_1Z(k)+\pi_{rs}E^T(k-\tau_s)E_2^TM^TPC_1E(k)\Big]\\
&+\sum_{s=1}^{q}\Big[\pi_{rs}Z^T(k-\tau_s)K_1^TN^TPNK_1Z(k-\tau_s)+\pi_{rs}Z^T(k-\tau_s)K_1^TN^TPMK_2E(k-\tau_s)\\
&+\pi_{rs}E^T(k-\tau_s)K_2^TM^TPNK_1Z(k-\tau_s)+\pi_{rs}E^T(k-\tau_s)K_2^TM^TPMK_2E(k-\tau_s)\Big]\\
&+Z^T(k)B_1^TPB_1Z(k)+Z^T(k)B_1^TPC_1E(k)+E^T(k)C_1^TPB_1Z(k)\\
&+E^T(k)C_1^TPC_1E(k)-Z^T(k)PZ(k)\\
&+\sum_{j=1}^{q}\left[Z^T(k)Q_jZ(k)-Z^T(k-\tau_j)Q_jZ(k-\tau_j)\right]\\
&+\sum_{j=1}^{q}\left[E^T(k)R_jE(k)-E^T(k-\tau_j)R_jE(k-\tau_j)\right]\\
&+\sum_{j=1}^{q}\eta^T(k)\tau_jZ_j\eta(k)-\sum_{j=1}^{q}\sum_{i=k-\tau_j}^{k-1}\eta^T(i)Z_j\eta(i)\\
&+2\sum_{j=1}^{q}\zeta^T(k)M_j\left[Z(k)-Z(k-\tau_j)-\sum_{l=k-\tau_j}^{k-1}\eta(l)\right]\\
&+\sum_{j=1}^{q}\sum_{i=k-\tau_j}^{k-1}\left[\zeta^T(k)M_j+\eta^T(k)Z_j\right]Z_j^{-1}\left[M_j^T\zeta(k)+Z_j\eta(k)\right]\\
&+Z^T(k)\overline{\Omega}Z(k)-E^T(k)\Omega Z(k)\\
&=\zeta^T(k)\left[\Phi_{11}(r)-\Phi_{21}^T\Phi_{22}^{-1}\Phi_{21}\right]\zeta(k),
\end{aligned}
$$

where $\Phi_{11}(r)$, Φ_{21} and Φ_{22} are given in (3.23), (3.24) and (3.25), respectively. To solve the consensus problem by using the Lyapunov functional method, the right-hand side of the above equation is needed to be less than zero to ensure that

$E\{\Delta V(k)\} < 0$. Let $S(r) \triangleq \Phi_{11}(r) - \Phi_{21}^T \Phi_{22}^{-1} \Phi_{21}$. Using the Schur complement, $S(r) < 0$ is equal to (3.22) holds. The subsequent aim is to show that the error dynamic system (3.13)–(3.16) subjects to random delays is stable in the mean square sense. Let the largest eigenvalue of $S(r)$ is $\lambda_{\max}(r)\ (< 0)$, we can obtain that

$$\begin{aligned} & E\{V(k+1) - V(k)\} \\ & \leq \zeta^T(k) S(r) \zeta(k) \\ & \leq \lambda_{\max}(r) \|\zeta(k)\|^2 \\ & \leq -\alpha \|\zeta(k)\|^2 \end{aligned} \tag{3.32}$$

with $\alpha = -\max\{\lambda_{\max}(r), r = 1, 2, \ldots, q\}$. Summing (3.32) from k=0, the following inequality can be derived

$$E\{V(k+1) - V(0)\} \leq -\alpha \sum_{l=0}^{k} E\left\{\|\zeta(l)\|^2\right\}.$$

Let $k \to \infty$, it can be obtained that

$$\begin{aligned} \sum_{l=0}^{k} E\left\{\|\zeta(l)\|^2\right\} & \leq \frac{1}{\alpha} E\{V(0) - V(k+1)\} \\ & \leq \frac{1}{\alpha} E\{V(0) - V(\infty)\} \\ & \leq \frac{1}{\alpha} E\{V(0)\}. \end{aligned}$$

Considering that $\|\zeta(l)\|^2 \geq \max\left\{\|x(l)\|^2, \|v(l)\|^2\right\}, l = 0, 1, 2, \ldots,$ therefore,

$$\begin{aligned} \sum_{l=0}^{k} E\left\{\|\delta(l)\|^2\right\} \leq \sum_{l=0}^{k} E\left\{\|\zeta(l)\|^2\right\} \leq \frac{1}{\alpha} E\{V(0)\}, \\ \sum_{l=0}^{k} E\left\{\|v(l)\|^2\right\} \leq \sum_{l=0}^{k} E\left\{\|\zeta(l)\|^2\right\} \leq \frac{1}{\alpha} E\{V(0)\}. \end{aligned}$$

According to the property of the convergent series, it can be concluded that

$$\lim_{l \to \infty} E\left\{\|\delta(l)\|^2\right\} = 0$$

and

$$\lim_{l \to \infty} E\left\{\|v(l)\|^2\right\} = 0.$$

Therefore, the error dynamic system (3.13)–(3.16) is mean square stable. So consensus is achieved in the mean square sensus according to Definition 3.1. This completes the proof.

Remark 3.4 The Lyapunov function used in the proof of Theorem 3.1 is constructed in light of [1]. As long as there is a feasible solution for the matrix inequalities (3.22) in Theorem 3.1, consensus can be guaranteed based on the event-triggered control protocols (3.9) and (3.10). However, the sufficient conditions in Theorem 3.1 is not in the form of LMIs which can be solved using the existing MATLAB Toolbox. In the following theorem, we will derive the sufficient conditions in the form of LMIs to grantee the consensus. Meanwhile, the explicit expression of the feedback gain matrix K in the event-triggered control protocols (3.9)–(3.10) is also obtained in the following Theorem 3.2.

Theorem 3.2 *Suppose that the communication topology G is directed and Assumption 3.1 and Assumption 3.2 hold, then the event-triggered control protocols (3.9) and (3.10) with given trigger parameters $\sigma_i (i \in S_n)$ in (3.7) can solve the consensus problem for the discrete-time heterogeneous MASs (3.1) and (3.2) with random delays governed by a Markov chain under Definition 3.1 if there exist symmetrical matrices $\widetilde{\mathcal{P}} > 0$ where $\widetilde{\mathcal{P}} \in R^{p \times p}$, $\widetilde{Q}_j > 0$, $\widetilde{R}_j > 0$, $\widetilde{Z}_j > 0$, $j = 1, 2, \ldots, q$, matrices $\widetilde{\Omega}_i > 0$, $i = 1, 2, \ldots, n$ and Y, $\widetilde{M}_j$, $j = 1, 2, \ldots, q$ with suitable dimensions such that the following LMIs are satisfied:*

$$
\begin{bmatrix}
\widetilde{\Theta}_0 + \widetilde{\Psi} + \widetilde{\Psi}^T & * & * & * & * & * & * & * \\
\widetilde{\Phi}_{21} & \widetilde{\Phi}_{22} & * & * & * & * & * & * \\
\Xi_1 & 0 & -\Lambda_1 & * & * & * & * & * \\
\Xi_2 & 0 & 0 & -\Lambda_2 & * & * & * & * \\
\Xi_3 & 0 & 0 & 0 & -\Lambda_3 & * & * & * \\
\Xi_4 & 0 & 0 & 0 & 0 & -\Lambda_3 & * & * \\
\Xi_5 & 0 & 0 & 0 & 0 & 0 & -\Lambda_4 & * \\
\Xi_6 & 0 & 0 & 0 & 0 & 0 & 0 & -\Lambda_4
\end{bmatrix} < 0 \tag{3.33}
$$

where

$$
\widetilde{\Theta}_0 = diag\left\{ -\widetilde{P}_1 + \widehat{\Omega} + \sum_{j=1}^{q} \widetilde{Q}_j,\ -\widetilde{Q}_1,\ \ldots,\ -\widetilde{Q}_q,\ \sum_{j=1}^{q} \widetilde{R}_j - \widetilde{\Omega},\ -\widetilde{R}_1,\ \ldots,\ -\widetilde{R}_q \right\},
$$

$$
\widetilde{\Psi} = \left[\sum_{j=1}^{q} \widetilde{M}_q\ -\widetilde{M}_1\ -\widetilde{M}_2\ \cdots\ -\widetilde{M}_q\ 0\ 0\ \cdots\ 0 \right],
$$

$$
\widetilde{\Phi}_{21} = \left[\sqrt{\tau_1}\widetilde{M}_1\ \sqrt{\tau_2}\widetilde{M}_2\ \cdots\ \sqrt{\tau_q}\widetilde{M}_q \right]^T,
$$

$$
\widetilde{\Phi}_{22} = -diag\left\{ \widetilde{Z}_1,\ \widetilde{Z}_2,\ \ldots,\ \widetilde{Z}_q \right\},
$$

$$
\Xi_1 = \begin{bmatrix}
\sqrt{\pi_{r1}}(B\widetilde{P}_1 + F_1Y_1) & \sqrt{\pi_{r1}}NY_1 & 0 & 0 & 0 & 0 & 0 & 0 \\
\vdots & 0 & \ddots & 0 & 0 & 0 & 0 & 0 \\
\sqrt{\pi_{rq}}(B\widetilde{P}_1 + F_1Y_1) & 0 & 0 & \sqrt{\pi_{rq}}NY_1 & 0 & 0 & 0 & 0 \\
BP_1 + F_1Y_1 & 0 & 0 & 0 & CP_2 + F_2Y_2 & 0 & 0 & 0 \\
\sqrt{\pi_{r1}}(B\widetilde{P}_1 + F_1Y_1) & 0 & 0 & 0 & 0 & \sqrt{\pi_{r1}}MY_2 & 0 & 0 \\
\vdots & 0 & 0 & 0 & 0 & 0 & \ddots & 0 \\
\sqrt{\pi_{rq}}(B\widetilde{P}_1 + F_1Y_1) & 0 & 0 & 0 & 0 & 0 & 0 & \sqrt{\pi_{rq}}MY_2
\end{bmatrix},
$$

$$\Xi_2 = \begin{bmatrix} \sqrt{\pi_{r1}}(B\widetilde{P}_1 - \widetilde{P}_1 + F_1Y_1) & \sqrt{\pi_{r1}}NY_1 & 0 & 0 & 0 & 0 & 0 & 0 \\ \vdots & 0 & \ddots & 0 & 0 & 0 & 0 & 0 \\ \sqrt{\pi_{rq}}(B\widetilde{P}_1 - \widetilde{P}_1 + F_1Y_1) & 0 & 0 & \sqrt{\pi_{rq}}NY_1 & 0 & 0 & 0 & 0 \\ B\widetilde{P}_1 - \widetilde{P}_1 + F_1Y_1 & 0 & 0 & 0 & C\widetilde{P}_2 + F_2Y_2 & 0 & 0 & 0 \\ \sqrt{\pi_{r1}}(B\widetilde{P}_1 - \widetilde{P}_1 + F_1Y_1) & 0 & 0 & 0 & 0 & \sqrt{\pi_{r1}}MY_2 & 0 & 0 \\ \vdots & 0 & 0 & 0 & 0 & 0 & \ddots & 0 \\ \sqrt{\pi_{rq}}(B\widetilde{P}_1 - \widetilde{P}_1 + F_1Y_1) & 0 & 0 & 0 & 0 & 0 & 0 & \sqrt{\pi_{rq}}MY_2 \end{bmatrix},$$

$$\Xi_3 = \begin{bmatrix} 0\,0\,0\,0 & \sqrt{\pi_{r1}}MY_2 & \sqrt{\pi_{r1}}MY_2 & 0 & \cdots & 0 \\ 0\,0\,0\,0 & \sqrt{\pi_{r2}}MY_2 & 0 & \sqrt{\pi_{r2}}MY_2 & \cdots & 0 \\ \vdots\,\vdots\,\vdots\,\vdots & \vdots & \vdots & \vdots & \ddots & \vdots \\ 0\,0\,0\,0 & \sqrt{\pi_{rq}}MY_2 & 0 & 0 & \cdots & \sqrt{\pi_{rq}}MY_2 \end{bmatrix},$$

$$\Xi_4 = \begin{bmatrix} 0 & \sqrt{\pi_{r1}}NY_1 & 0 & \cdots & 0 & 0\,0\,0\,0 \\ 0 & 0 & \sqrt{\pi_{r2}}NY_1 & \cdots & 0 & 0\,0\,0\,0 \\ \vdots & \vdots & \vdots & \ddots & \vdots & \vdots\,\vdots\,\vdots\,\vdots \\ 0 & 0 & 0 & \cdots & \sqrt{\pi_{rq}}NY_1 & 0\,0\,0\,0 \end{bmatrix},$$

$$\Xi_5 = \begin{bmatrix} 0\,0\,0\,0\,0 & 0 & 0 & \cdots & 0 \\ 0\,0\,0\,0\,0 & \sqrt{\pi_{r1}}MY_2 & 0 & \cdots & 0 \\ 0\,0\,0\,0\,0 & 0 & \sqrt{\pi_{r2}}MY_2 & \cdots & 0 \\ \vdots\,\vdots\,\vdots\,\vdots\,\vdots & \vdots & \vdots & \ddots & \vdots \\ 0\,0\,0\,0\,0 & 0 & 0 & \cdots & \sqrt{\pi_{rq}}MY_2 \end{bmatrix},$$

$$\Xi_6 = \begin{bmatrix} 0\,0\,0\,0\,0 & 0 & 0 & \cdots & 0 \\ 0\,0\,0\,0\,0 & \sqrt{\pi_{r1}}\left(C\widetilde{P}_2 + F_2Y_2\right) & 0 & \cdots & 0 \\ 0\,0\,0\,0\,0 & 0 & \sqrt{\pi_{r2}}\left(C\widetilde{P}_2 + F_2Y_2\right) & \cdots & 0 \\ \vdots\,\vdots\,\vdots\,\vdots\,\vdots & \vdots & \vdots & \ddots & \vdots \\ 0\,0\,0\,0\,0 & 0 & 0 & \cdots & \sqrt{\pi_{rq}}\left(C\widetilde{P}_2 + F_2Y_2\right) \end{bmatrix},$$

$$\begin{aligned} \Lambda_1 &= diag\left\{\widetilde{P}_1, \ldots, \widetilde{P}_1, \widetilde{P}_1, \widetilde{P}_1, \ldots, \widetilde{P}_1\right\}, \\ \Lambda_2 &= diag\left\{\widehat{Z} - 2\widetilde{P}_1, \ldots, \widehat{Z} - 2\widetilde{P}_1, \widehat{Z} - 2\widetilde{P}_1, \widehat{Z} - 2\widetilde{P}_1, \ldots, \widehat{Z} - 2\widetilde{P}_1\right\}, \\ \Lambda_3 &= diag\left\{\widehat{Z} - \widetilde{P}_1, \widehat{Z} - \widetilde{P}_1, \ldots, \widehat{Z} - \widetilde{P}_1\right\}, \\ \Lambda_4 &= diag\left\{\widehat{Z} - \widetilde{P}_1, \widehat{Z} - \widetilde{P}_1, \widehat{Z} - \widetilde{P}_1, \ldots, \widehat{Z} - \widetilde{P}_1\right\}, \\ \widehat{Z} &= \sum_{j=1}^{q} \tau_j \widetilde{Z}_j, \\ \widetilde{P}_1 &= I_{(n+l)} \otimes \widetilde{\mathcal{P}}, \widetilde{P}_2 = I_n \otimes \widetilde{\mathcal{P}}, \\ Y_1 &= K_1\widetilde{P}_1 = I_{(n+l)} \otimes Y, Y_2 = K_2\widetilde{P}_2 = I_n \otimes Y \end{aligned} \tag{3.34}$$

and

$$\widehat{\Omega} = diag\left\{0, \ldots, 0, \sigma_1\widetilde{\Omega}_1, \ldots, \sigma_l\widetilde{\Omega}_l, \sigma_{l+1}\widetilde{\Omega}_{l+1}, \ldots, \sigma_n\widetilde{\Omega}_n\right\},$$
$$\widetilde{\Omega} = diag\left\{\widetilde{\Omega}_1, \widetilde{\Omega}_2, \ldots, \widetilde{\Omega}_l, \widetilde{\Omega}_{l+1}, \ldots, \widetilde{\Omega}_n\right\}.$$

Furthermore, the feedback gain matrix K in (3.9) and (3.10) and the trigger matrices Ω_i in (3.7) are given as $K = Y\widetilde{\mathcal{P}}^{-1}$, $\Omega_i = \widetilde{\mathcal{P}}^{-1}\widetilde{\Omega}_i\widetilde{\mathcal{P}}^{-1}$, $i \in S_n$.

Proof For simplicity and further analysis, the positive definite matrix P in Theorem 3.1 is selected as $P = \left(I_{(n+l)} \otimes \mathcal{P}\right) \in R^{(l+n)p \times (l+n)p}$. Defining

$$P_1 = I_{(n+l)} \otimes \mathcal{P},$$
$$P_2 = I_n \otimes \mathcal{P},$$

and taking into account K_1 and K_2 in (3.21), $\Theta(r)$ in (3.23) can be rewritten as

$$\Theta(r) = \Theta_1 + \digamma_1^T\Upsilon_1\digamma_1 + \digamma_2^T\Upsilon_2\digamma_2 + \digamma_3^T\Upsilon_3\digamma_4 + s\digamma_4^T\Upsilon_3\digamma_3 + \digamma_5^T\Upsilon_4\digamma_6 + \digamma_6^T\Upsilon_4\digamma_5 \tag{3.35}$$

where

$$\begin{aligned}
\Theta_1 &= diag\left\{-P_1 + \overline{\Omega} + \sum_{j=1}^{q} Q_j - 2(BP_1 + F_1K_1)^T(BP_1 + F_1K_1) - 2(BP_1 - P_1 + F_1K_1)^T(BP_1 \right. \\
&\quad \left. - P_1 + F_1K_1), -Q_1, \ldots, -Q_q, \sum_{j=1}^{q} R_j - \Omega, -R_1, \ldots, -R_q\right\} \\
&\le diag\left\{-P_1 + \overline{\Omega} + \sum_{j=1}^{q} Q_j, -Q_1, \ldots, -Q_q, \sum_{j=1}^{q} R_j - \Omega, -R_1, \ldots, -R_q\right\} \\
&\triangleq \Theta_0,
\end{aligned} \tag{3.36}$$

$$\digamma_1 = \begin{bmatrix}
\sqrt{\pi_{r1}}(BP_1 + F_1K_1) & \sqrt{\pi_{r1}}NK_1 & 0 & 0 & 0 & 0 & 0 & 0 \\
\vdots & 0 & \ddots & 0 & 0 & 0 & 0 & 0 \\
\sqrt{\pi_{rq}}(BP_1 + F_1K_1) & 0 & 0 & \sqrt{\pi_{rq}}NK_1 & 0 & 0 & 0 & 0 \\
BP_1 + F_1K_1 & 0 & 0 & 0 & CP_2 + F_2K_2 & 0 & 0 & 0 \\
\sqrt{\pi_{r1}}(BP_1 + F_1K_1) & 0 & 0 & 0 & 0 & \sqrt{\pi_{r1}}MK_2 & 0 & 0 \\
\vdots & 0 & 0 & 0 & 0 & 0 & \ddots & 0 \\
\sqrt{\pi_{rq}}(BP_1 + F_1K_1) & 0 & 0 & 0 & 0 & 0 & 0 & \sqrt{\pi_{rq}}MK_2
\end{bmatrix},$$

$$\digamma_2 = \begin{bmatrix}
\sqrt{\pi_{r1}}(BP_1 - P_1 + F_1K_1) & \sqrt{\pi_{r1}}NK_1 & 0 & 0 & 0 & 0 & 0 & 0 \\
\vdots & 0 & \ddots & 0 & 0 & 0 & 0 & 0 \\
\sqrt{\pi_{rq}}(BP_1 - P_1 + F_1K_1) & 0 & 0 & \sqrt{\pi_{rq}}NK_1 & 0 & 0 & 0 & 0 \\
BP_1 - P_1 + F_1K_1 & 0 & 0 & 0 & CP_2 + F_2K_2 & 0 & 0 & 0 \\
\sqrt{\pi_{r1}}(BP_1 - P_1 + F_1K_1) & 0 & 0 & 0 & 0 & \sqrt{\pi_{r1}}MK_2 & 0 & 0 \\
\vdots & 0 & 0 & 0 & 0 & 0 & \ddots & 0 \\
\sqrt{\pi_{rq}}(BP_1 - P_1 + F_1K_1) & 0 & 0 & 0 & 0 & 0 & 0 & \sqrt{\pi_{rq}}MK_2
\end{bmatrix},$$

$$F_3 = \begin{bmatrix} 0\,0\,0\,0 & \sqrt{\pi_{r1}}MK_2 & \sqrt{\pi_{r1}}MK_2 & 0 & \cdots & 0 \\ 0\,0\,0\,0 & \sqrt{\pi_{r2}}MK_2 & 0 & \sqrt{\pi_{r2}}MK_2 & \cdots & 0 \\ \vdots\,\vdots\,\vdots\,\vdots & \vdots & \vdots & \vdots & \ddots & \vdots \\ 0\,0\,0\,0 & \sqrt{\pi_{rq}}MK_2 & 0 & 0 & \cdots & \sqrt{\pi_{rq}}MK_2 \end{bmatrix},$$

$$F_4 = \begin{bmatrix} 0 & \sqrt{\pi_{r1}}NK_1 & 0 & \cdots & 0 & 0\,0\,0\,0 \\ 0 & 0 & \sqrt{\pi_{r2}}NK_1 & \cdots & 0 & 0\,0\,0\,0 \\ \vdots & \vdots & \vdots & \ddots & \vdots & \vdots\,\vdots\,\vdots\,\vdots \\ 0 & 0 & 0 & \cdots & \sqrt{\pi_{rq}}NK_1 & 0\,0\,0\,0 \end{bmatrix},$$

$$F_5 = \begin{bmatrix} 0\,0\,0\,0\,0 & 0 & 0 & \cdots & 0 \\ 0\,0\,0\,0\,0 & \sqrt{\pi_{r1}}MK_2 & 0 & \cdots & 0 \\ 0\,0\,0\,0\,0 & 0 & \sqrt{\pi_{r2}}MK_2 & \cdots & 0 \\ \vdots\,\vdots\,\vdots\,\vdots\,\vdots & \vdots & \vdots & \ddots & \vdots \\ 0\,0\,0\,0\,0 & 0 & 0 & \cdots & \sqrt{\pi_{rq}}MK_2 \end{bmatrix},$$

$$F_6 = \begin{bmatrix} 0\,0\,0\,0\,0 & 0 & 0 & \cdots & 0 \\ 0\,0\,0\,0\,0 & \sqrt{\pi_{r1}}(CP_2 + F_2K_2) & 0 & \cdots & 0 \\ 0\,0\,0\,0\,0 & 0 & \sqrt{\pi_{r2}}(CP_2 + F_2K_2) & \cdots & 0 \\ \vdots\,\vdots\,\vdots\,\vdots\,\vdots & \vdots & \vdots & \ddots & \vdots \\ 0\,0\,0\,0\,0 & 0 & 0 & \cdots & \sqrt{\pi_{rq}}(CP_2 + F_2K_2) \end{bmatrix},$$

$$\Upsilon_1 = diag\left\{ P_1, \ldots, P_1, P_1, P_1, \ldots, P_1 \right\},$$

$$\Upsilon_2 = diag\left\{ \overline{Z}, \ldots, \overline{Z}, \overline{Z}, \overline{Z}, \ldots, \overline{Z} \right\}, \tag{3.37}$$

$$\Upsilon_3 = diag\left\{ P_1 + \overline{Z}, P_1 + \overline{Z}, \ldots, P_1 + \overline{Z} \right\}, \tag{3.38}$$

$$\Upsilon_4 = diag\left\{ P_1 + \overline{Z}, P_1 + \overline{Z}, P_1 + \overline{Z}, \ldots, P_1 + \overline{Z} \right\}, \tag{3.39}$$

According to Lemma 3.2, for any matrices X and Y, if $P > 0$, the following inequality holds

$$X^T PY + Y^T PX \leq X^T PX + Y^T PY. \tag{3.40}$$

By combining (3.35), (3.36) and using the inequality (3.40), the condition in (3.22) in Theorem 3.1 satisfies

$$\begin{bmatrix} \Theta(r) + \Psi + \Psi^T & * \\ \Phi_{21} & \Phi_{22} \end{bmatrix} \leq \begin{bmatrix} \Theta_0 + \Psi + \Psi^T & * \\ \Phi_{21} & \Phi_{22} \end{bmatrix} + \Pi \Upsilon \Pi^T \tag{3.41}$$

where

$$\Pi = \begin{bmatrix} F_1^T & F_2^T & F_3^T & F_4^T & F_5^T & F_6^T \\ 0 & 0 & 0 & 0 & 0 & 0 \end{bmatrix},$$

$$\Upsilon = diag\left\{ \Upsilon_1, \Upsilon_2, \Upsilon_3, \Upsilon_3, \Upsilon_4, \Upsilon_4 \right\}.$$

Furthermore, using the Shur complement, the above inequality (3.41) is equal to

$$\begin{bmatrix} \Theta_0+\Psi+\Psi^T & * & * & * & * & * & * & * \\ \Phi_{21} & \Phi_{22} & * & * & * & * & * & * \\ F_1 & 0 & -\Upsilon_1^{-1} & * & * & * & * & * \\ F_2 & 0 & 0 & -\Upsilon_2^{-1} & * & * & * & * \\ F_3 & 0 & 0 & 0 & -\Upsilon_3^{-1} & * & * & * \\ F_4 & 0 & 0 & 0 & 0 & -\Upsilon_3^{-1} & * & * \\ F_5 & 0 & 0 & 0 & 0 & 0 & -\Upsilon_4^{-1} & * \\ F_6 & 0 & 0 & 0 & 0 & 0 & 0 & -\Upsilon_4^{-1} \end{bmatrix} < 0.$$

According to (3.37)–(3.39), the terms Υ_2^{-1}, Υ_3^{-1} and Υ_4^{-1} in the above inequality are difficult to deal with using the existing numerical software. So, some more manipulation is needed. Pre and post-multiplying it with

$$diag\left\{ I_{q+1}\otimes P_1^{-1},\ I_{q+1}\otimes P_2^{-1},\ I_q\otimes P_1^{-1},\ I_{2q+1}\otimes I_{(n+l)p},\ I_{2q+1}\otimes I_{(n+l)p},\ I_{2q+1}\otimes I_{(n+l)p},\ I_{2q+1}\otimes I_{(n+l)p},\ I_{2q+1}\otimes I_{(n+l)p},\ I_{2q+1}\otimes I_{(n+l)p} \right\},$$

where I_m represent the identity matrix with dimension m. Defining

$$\begin{aligned} &\widetilde{\mathcal{P}}=\mathcal{P}^{-1},\ \widetilde{P}_1=P_1^{-1},\ \widetilde{P}_2=P_2^{-1},\\ &\widetilde{Q}_i=\widetilde{P}_1Q_i\widetilde{P}_1,\ \widetilde{R}_i=\widetilde{P}_2R_i\widetilde{P}_2,\ \widetilde{M}_i=\widetilde{P}_1M_i\widetilde{P}_1,\\ &\widetilde{Z}_i=\widetilde{P}_1Z_i\widetilde{P}_1,\ i=1,2,\ldots,q,\\ &\widetilde{\Omega}_i=\widetilde{\mathcal{P}}\Omega_i\widetilde{\mathcal{P}},\ i=1,2,\ldots,n,\\ &Y=K\widetilde{\mathcal{P}},\ Y_1=K\widetilde{P}_1,\ Y_2=K\widetilde{P}_2, \end{aligned} \tag{3.42}$$

then the following inequality is obtained:

$$\begin{bmatrix} \widetilde{\Theta}_0+\widetilde{\Psi}+\widetilde{\Psi}^T & * & * & * & * & * & * & * \\ \widetilde{\Phi}_{21} & \widetilde{\Phi}_{22} & * & * & * & * & * & * \\ \Xi_1 & 0 & -\Lambda_1 & * & * & * & * & * \\ \Xi_2 & 0 & 0 & -\Upsilon_2^{-1} & * & * & * & * \\ \Xi_3 & 0 & 0 & 0 & -\Upsilon_3^{-1} & * & * & * \\ \Xi_4 & 0 & 0 & 0 & 0 & -\Upsilon_3^{-1} & * & * \\ \Xi_5 & 0 & 0 & 0 & 0 & 0 & -\Upsilon_4^{-1} & * \\ \Xi_6 & 0 & 0 & 0 & 0 & 0 & 0 & -\Upsilon_4^{-1} \end{bmatrix} < 0 \tag{3.43}$$

where $\widetilde{\Theta}_0$, $\widetilde{\Psi}$, $\widetilde{\Phi}_{21}$, $\widetilde{\Phi}_{22}$, Λ_1, Ξ_i, $i=1,2,\ldots,6$ are as the same statement in Theorem 3.2. For any matrix $Z>0$, it is well known that

$$-PZ^{-1}P\le Z-2P.$$

Considering the above inequality (3.43) and taking into account (3.26), (3.34), (3.37) and (3.42), the following inequality can be derived

$$\begin{aligned} -\overline{Z}^{-1} &= \widetilde{P}_1 \left(-P_1 \overline{Z}^{-1} P_1\right) \widetilde{P}_1 \\ &\leq \widetilde{P}_1 \left(\overline{Z} - 2P_1\right) \widetilde{P}_1 \\ &= \widehat{\overline{Z}} - 2\widetilde{P}_1, \end{aligned}$$

from which it can be easily seen that $-\Upsilon_2^{-1} \leq \Lambda_2$. Using the same technique, the following inequalities $\Upsilon_3^{-1} \leq \Lambda_3$ and $\Upsilon_4^{-1} \leq \Lambda_4$ hold. Therefore, if the LMIs in (3.33) hold for all $r = 1, 2, \ldots, q$, the error dynamic system (3.13)–(3.16) is mean square stable, and the consensus can be guaranteed. This completes the proof.

3.4 Simulation Example

To illustrate the obtained theoretical results above, in this section we consider a system consisting of four agents indexed by 1, 2, 3, 4, respectively, which is categorized by the following two subsystems: the first two agents are described by the following linear second-order dynamics

$$\begin{aligned} x_i(k+1) &= x_i(k) + v_i(k) \\ v_i(k+1) &= v_i(k) + u_i(k), i = 1, 2 \end{aligned}$$

where $u_i(k)$ is defined in (3.9) and the last two agents are described by the following linear first-order dynamics

$$x_i(k+1) = x_i(k) + u_i(k), i = 3, 4$$

where $u_i(k)$ is defined in (3.10).

The communication graph G which has a spanning tree for the four agents is shown in Fig. 3.1. The adjacent matrix is chosen as $A = \begin{bmatrix} 0 & 0.2 & 0 & 0 \\ 0.4 & 0 & 0.17 & 0 \\ 0.24 & 0 & 0 & 0 \\ 0 & 0.2 & 0.18 & 0 \end{bmatrix}$, so the Laplacian matrix $L = \begin{bmatrix} 0.2 & -0.2 & 0 & 0 \\ -0.4 & 0.57 & -0.17 & 0 \\ -0.24 & 0 & 0.24 & 0 \\ 0 & -0.2 & -0.18 & 0.38 \end{bmatrix}$. H is chosen as $H = I$. Without loss of generality, we let $b_0 = \mathbf{0}$. Set the constant feedback gain γ as $\gamma = 0.8$. The delay set Γ is chosen as $\Gamma = \{3, 5, 9\}$ and its transition probability matrix is

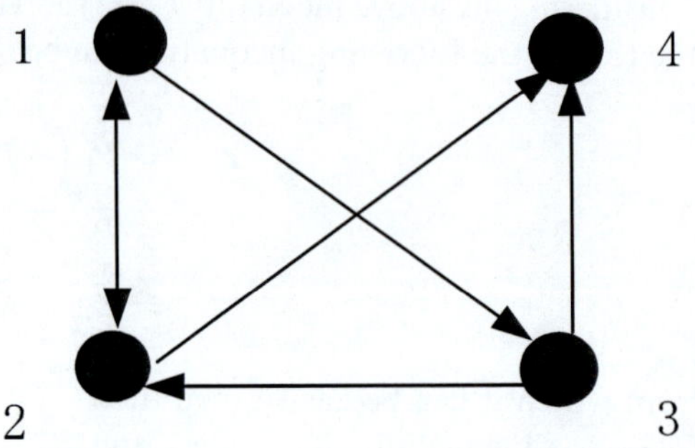

Fig. 3.1 Directed communication graph G

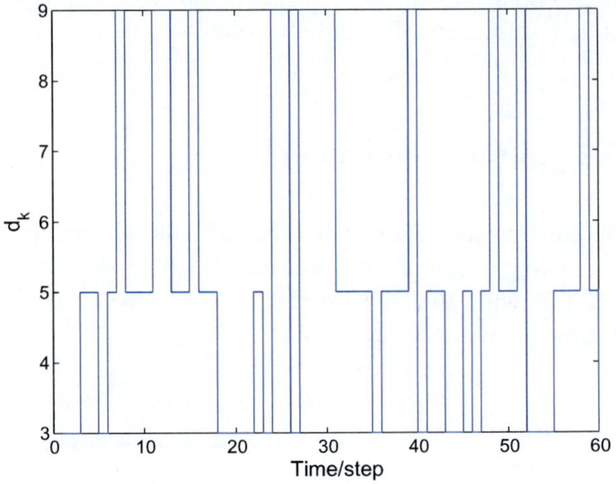

Fig. 3.2 Time delay (d_k) over time

$$\Pi = \begin{bmatrix} 0.3 & 0.5 & 0.2 \\ 0.6 & 0.3 & 0.1 \\ 0.4 & 0.4 & 0.2 \end{bmatrix}.$$

The trigger conditions (3.6) and (3.7) are applied to the event-triggered control strategy. The trigger parameters σ_i $(i = 1, 2, 3, 4)$ are chosen as $\sigma_1 = 0.2, \sigma_2 = 0.1, \sigma_1 = 0.3, \sigma_2 = 0.15$. A feasible solution to (3.33) in Theorem 3.2 can be shown as follows: the feedback gain matrix K is $K = diag\left\{0.4806 \quad 0.3720\right\}$ and the trigger matrices are

$$\Omega_1 = \begin{bmatrix} 0.1007 & 0.0032 \\ 0.0032 & 0.1096 \end{bmatrix}, \Omega_2 = \begin{bmatrix} 0.4085 & 0.0124 \\ 0.0124 & 0.2513 \end{bmatrix},$$
$$\Omega_3 = \begin{bmatrix} 0.2350 & -0.0074 \\ -0.0074 & 0.1035 \end{bmatrix}, \Omega_4 = \begin{bmatrix} 0.1581 & -0.0004 \\ -0.0004 & 0.0467 \end{bmatrix}.$$

Time jumping delay over time is given in Fig. 3.2. Based on the feedback matrix K and trigger matrices $\Omega_i, i = 1, 2, 3, 4$ obtained above and the initial

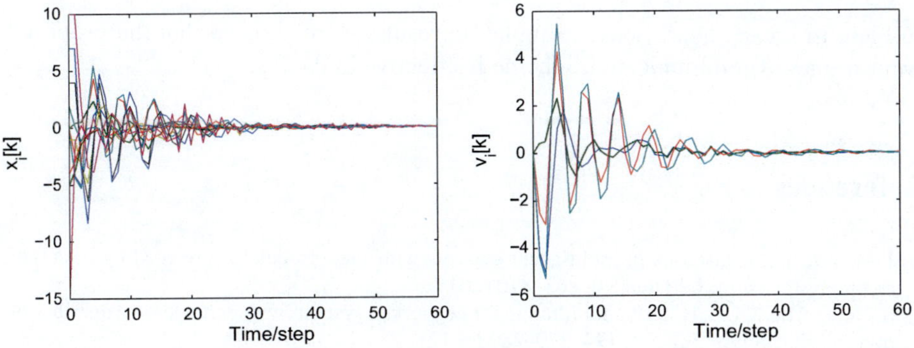

Fig. 3.3 The evolution of four agents :**a** State; **b** Velocity

Table 3.1 Sampled and released times for states

	Agent 1	Agent 2	Agent 3	Agent 4
Sampled times	60	60	60	60
Released times	51	38	47	52

states and initial velocities are chosen as $x_1(0) = [3, -1]^T$, $x_2(0) = [4, 7]^T$, $v_1(0) = [0, 0]^T$, $v_2(0) = [0, 0]^T$, $x_3(0) = [-5, -4]^T$, $x_4(0) = [2, -15]^T$, the states evolution of the four agents are drawn in Fig. 3.3a and the velocities evolution of the first two agents are drawn in Fig. 3.3b, respectively. The sampled and event numbers for states of agents is given in Table 3.1.

It is observed from Fig. 3.3 that all agents reach consensus within 60 steps. By simple computation according to Table 3.1, for $k \in [0, 60]$, only 78.33% of all sampled signals of all agents about state are needed to be sent out to their neighbors. So the proposed event-triggered scheme can reduce much signal transmission and thus reduce the multi-agent network communication burden than the traditional communication scheme while preserve the desired consensus performance. This demonstrates the effectiveness of the proposed consensus algorithm in this chapter.

3.5 Conclusion

In order to save the communication resources and reduce the communication burden for the multi-agent network, in this chapter, the distributed event-triggered consensus controls were proposed for discrete-time heterogeneous MASs with random communication delays represented by a homogeneous Markov chain. The interaction graph was assumed to be directed and have a directed spanning tree. Sufficient conditions in forms of linear matrix inequalities have been derived by using the Lyapunov functional method and the Kronecker product technology to solve the consensus problem in mean square sense. Simulation results were showed that the proposed event-triggered communication scheme is effective in the last.

References

1. J. Wu, Y. Shi, Consensus in multi-agent systems with random delays governed by a markov chain. Syst. Control Lett. **60**(10), 863–870 (2011)
2. S. Hu, D. Yue, J. Liu, Hiltering for filtering for networked systems with partly known distribution transmission delays. Inf. Sci. **194**, 270–282 (2012)
3. Y.G. Sun, L. Wang, Consensus of multi-agent systems in directed networks with nonuniform time-varying delays. IEEE Trans. Autom. Control **54**(7), 1607–1613 (2009)
4. Y. Hu, H. Su, J. Lam, Adaptive consensus with a virtual leader of multiple agents governed by locally lipschitz nonlinearity. Int. J. Robust Nonlinear Control **23**(9), 978–990 (2013)
5. J.-S. Chiou, C.-J. Wang, C.-M. Cheng, On delay-dependent stabilization analysis for the switched time-delay systems with the state-driven switching strategy. J. Franklin Inst. **348**(2), 261–276 (2011)
6. W. Yu, W.X. Zheng, G. Chen, W. Ren, J. Cao, Second-order consensus in multi-agent dynamical systems with sampled position data. Automatica **47**(7), 1496–1503 (2011)
7. H. Zhang, D. Yue, C. Dou, W. Zhao, X. Xie, Data-Driven distributed optimal consensus control for unknown multiagent systems with input-delay. IEEE Trans. Cybern. **49**(6), 2095–2105 (2019)

Part II
Distributed Cooperative Control of Nonlinear Multi-agent Systems

Chapter 4
Distributed Formation Tracking Conrol of Nonlinear Mutiagent Systems

This chapter addresses the time-varying formation tracking issue for a class of uncertain non-affine nonlinear MASs with communication delays. The agents are connected though a directed topology, and the non-affine problem is transformed into affine problem using the idea of active disturbance rejection control (ADRC). The uncertainties of individual followers in the MAS are approximated via the extended state observers (ESOs) and are compensated in real time. Tracking differentiators (TDs) are introduced to overcome the explosion of complexity in the derivatives of the desired signal and the virtual control variables. Local information from neighboring followers are utilized, and the Lyapunov-Krasovskii function is employed to analyse the stability theoretically. Finally, simulation examples are provided to demonstrate the effectiveness of the proposed control strategy.

This chapter is structured as below, in Sect. 4.1, the system model of this chapter is described, and the distributed secure consensus protocol is designed. Section 4.2 presents the stability analysis of the proposed scheme and explained the switching topology. In Sect. 4.3, the simulation results of numerical example and a group of unmanned aerial vehicles (UAVs) are given to prove the feasibility of the distributed secure consensus protocol. Finally, Sect. 4.4 concludes the chapter.

4.1 Problem Formulation

Consider a class of nonlinear MASs, and the system is composed of one leader and N followers. The dynamics of the ith follower are

D. Yue et al., *Distributed Cooperative Control and Communication for Multi-agent Systems*, https://doi.org/10.1007/978-981-33-6718-0_4

$$\begin{cases} \dot{x}_{i,1}(t) = f_{i,1}(x_{i,1}, x_{i,2}), \\ \dot{x}_{i,2}(t) = f_{i,2}(\bar{x}_{i,2}, x_{i,3}), \\ \quad \vdots \\ \dot{x}_{i,n_i}(t) = f_{i,n_i}(\bar{x}_{i,n_i}, u_i), \\ y_i = x_{i,1}, \end{cases} \tag{4.1}$$

where $\bar{x}_{i,p} = [x_{i,1}, x_{i,2}, \ldots, x_{i,p}]^T \in R^p$ is a state vector and is available for measurement, $i = 1, \ldots, N$, $p = 1, \ldots, n_i$, $y = [y_1, \ldots, y_N]^T \in R^N$ are the output signals of the followers, $u_i \in R$ is the control input of the ith follower, $f_{i,p}(\cdot) : R^{p+1} \to R$ is an unknown continuous function containing uncertainties of the ith follower. It is assumed that the signal of the leader y_d is a constant. Define the following partial derivatives $g_{i,p}\left(\bar{x}_{i,p}, x_{i,p+1}\right) = \partial f_{i,p}\left(\bar{x}_{i,p}, x_{i,p+1}\right) / \partial x_{i,p+1}$ and $g_{i,n_i}\left(\bar{x}_{i,n_i}, u_i\right) = \partial f_{i,n_i}\left(\bar{x}_{i,n_i}, u_i\right) / \partial u_i$, for $i = 1, \ldots, N$, $p = 1, \ldots, n_i - 1$. To move on, the following assumption is made.

Assumption 4.1 ([1, 2]) There exist unknown positive constants $g_{i,j,0}$ and $g_{i,j,1}$, such that $0 < g_{i,j,0} < |g_{i,j}(\cdot)| < g_{i,j,1} < \infty$.

Assumption 4.1 indicates that $g_{i,j}(\cdot) \neq 0$. Without loss of generality, we mainly consider the scenario that $0 < g_{i,j,0} < g_{i,j}(\cdot) < g_{i,j,1} < \infty$ in this chapter. Assumption 4.1 is fairly mild and common in the existing references, such as [2].

Definition 4.1 ([3, 4]) If there exist a continuous function $V \in C^1$, class $\mathcal{K}$ functions $\kappa_1(s)$, $\kappa_2(s)$, $\kappa_3(s)$, $\kappa_4(s)$, and a non-negative constant d such that

$$\begin{cases} \kappa_1(||x||) \leq V(x) \leq \kappa_2(||x||), \forall x \in \mathbb{R}^n, \\ \frac{\partial V}{\partial x} f(x, u) \leq -\kappa_3(||x||) + \kappa_4(||u||) + d, \end{cases} \tag{4.2}$$

$V(x)$ is the input-to-state practically stability (ISpS) function.

Assumption 4.2 ([5, 6]) The directed communication graph $\mathcal{G}$ is fixed and strongly connected, and at least one follower in $\mathcal{G}$ has direct access to y_d.

Lemma 4.1 ([7]) *For a directed graph satisfying Assumption 4.2, there exists a positive vector $w = [w_1, \ldots, w_N]^T$ such that $w^T \mathcal{L} = 0$ if w_i is the left eigenvector of $\mathcal{L}$ associated with $\lambda(\mathcal{L}) = 0$, where $\lambda(\star)$ is the eigenvalue of the matrix $\star$.*

In this chapter, we aim for, with unknown constant communication delays among followers, designing a formation tracking control strategy to force the outputs of each follower y_i and the signal of the leader y_d in the desired formation shape. As the equation $y_i - \delta_i = y_d$ holds, the MAS realizes the desired formation shape, where δ_i represents the assigned outputs of the ith follower in the desired formation shape and is available to its neighbors. In other words, we are going to develop a formation tracking control strategy u_i such that the followers can track the output signal of leader y_d despite of communication delays among followers while the formation tracking

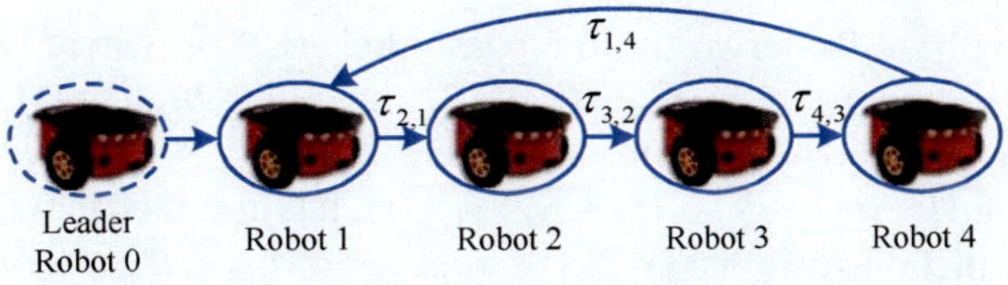

Fig. 4.1 The communication graph for a team of mobile robots with delays

error $y_i - \delta_i - y_d$ is in a residue around the origin, $i = 1, \ldots, N$. One motivating example, a team of mobile robots, for the problem is presented in Fig. 4.1, and more details can be found in Sect. 4.3.

4.2 Formation Tracking Control Strategy Design

We are going to derive a time-varying formation tracking strategy for the follower (4.1), and the stability of the closed-loop system will be discussed.

From Assumption 4.1, (4.1) is rewritten as

$$\begin{cases} \dot{x}_{i,1}(t) = F_{i,1}(x_{i,1}, x_{i,2}) + b_{i,1}x_{i,2}, \\ \dot{x}_{i,2}(t) = F_{i,2}(\bar{x}_{i,2}, x_{i,3}) + b_{i,2}x_{i,3}, \\ \qquad \vdots \\ \dot{x}_{i,n_i}(t) = F_{i,n_i}(\bar{x}_{i,n_i}, u_i) + b_{i,n_i}u_i, \\ y_i = x_{i,1}, \end{cases} \tag{4.3}$$

where $i = 1, \ldots, N$, $b_{i,p}$ is a positive constant, $p = 1, \ldots, n_i$, and its symbol are the same as $g_{i,p}\left(\bar{x}_{i,p}, x_{i,p+1}\right)$ and $\partial f_{n_i}\left(\bar{x}_{i,n_i}, u_i\right)$ in Assumption 4.1. The unknown function $F_{i,p}(\cdot)$ is defined as $F_{i,p}(\bar{x}_{i,p}, x_{i,p+1}) = f_{i,p}(\bar{x}_{i,p}, x_{i,p+1}) - b_{i,p}x_{i,p+1}$, $1 \leq p \leq n_i - 1$, and $F_{i,n_i}(\bar{x}_{i,n_i}, u_i) = f_{i,n_i}(\bar{x}_{i,n_i}, u_i) - b_{i,n_i}u_i$. It should be pointed out that, in the framework of backstepping method, a direct design controller for the system (4.3) may not be feasible. This is due to the fact that the state $x_{i,p+1}$ is used in the design of virtual control variable $\alpha_{i,p+1}$, $1 \leq p \leq n_i - 1$, and it is called an algebraic circle. To address this issue, we transform (4.3) into

$$\begin{cases} \dot{x}_{i,1}(t) = \bar{F}_{i,1}(t) + b_{i,1}x_{i,2,f}, \\ \dot{x}_{i,2}(t) = \bar{F}_{i,2}(t) + b_{i,2}x_{i,3,f}, \\ \qquad \vdots \\ \dot{x}_{i,n_i}(t) = \bar{F}_{i,n_i}(t) + b_{i,n_i}u_{i,f}, \\ y_i = x_{i,1}, \end{cases} \tag{4.4}$$

with a Butterworth low-pass filter (LPF), where $\bar{F}_{i,p}(t) = F_{i,p}(\bar{x}_{i,p}, x_{i,p+1}) + b_{i,p}l_{i,p}(t)$, $\bar{F}_{i,n_i}(t) = F_{i,n_i}(\bar{x}_{i,n_i}, u_{i,p+1}) + b_{i,n_i}l_{i,n_i}(t)$, $x_{i,p+1,f} = H_L(s)x_{i,p+1} \approx x_{i,p+1}$, $u_{i,f} = H_L(s)u_i \approx u_i$, $p = 1, \ldots, n_i - 1$, $H_L(s)$ is a Butterworth LPF [8, 9], and $l_{i,p}(t) = x_{i,p+1}(t) - x_{i,p+1,f}(t)$. It is assumed that $l_{i,p}(t)$ is an unknown bounded function such that $|x_{i,p+1} - x_{i,p+1,f}| \leq l_{i,p,0}$, $l_{i,p,0} > 0$.

For the convenience of design of the control strategy, we define the following errors

$$\begin{cases} e_{i,1}(t) = y_i - c_i y_d - (1 - c_i)\hat{y}_{di} - \delta_i(t), \\ e_{i,j}(t) = x_{i,j} - \alpha_{i,j}(t), \quad j = 2, \ldots, n_i, \end{cases} \tag{4.5}$$

where $\hat{y}_{di}$ is the estimation of y_d introduced for the ith follower if $c_i = 0$, and $\alpha_{i,j}(t)$ is the virtual control variable. If the ith follower can access to the leader's signal y_d directly, $c_i = 1$, otherwise c_i set to 0.

In the following part, we will design the time-varying formation control strategy for the ith follower (4.1) with constant communication delay via the ADRC, backstepping technique and the Lyapunov-Krasovskii function. The block diagram for the ith follower's control strategy is plotted in Fig. 4.2.

Step i, 1: The derivative of $e_{i,1}$ is

$$\dot{e}_{i,1}(t) = \bar{F}_{i,1} + b_{i,1}x_{i,2,f} - c_i\dot{y}_d - (1 - c_i)\dot{\hat{y}}_{di} - \dot{\delta}_i(t). \tag{4.6}$$

Due to the existence of the unknown item $\bar{F}_{i,1}(\cdot)$, an ESO

$$\begin{cases} E_{i,1} = \phi_{i,1,1} - e_{i,1}, \\ \dot{\phi}_{i,1,1} = \phi_{i,1,2} - \rho_{i,1,1}E_{i,1} + b_{i,1}x_{i,2,f} - (1 - c_i)\dot{\hat{y}}_{di} - \dot{\delta}_i, \\ \dot{\phi}_{i,1,2} = -\rho_{i,1,2}|E_{i,1}|^{\sigma_{i,1,2}}\text{sign}(E_{i,1}), \end{cases} \tag{4.7}$$

is employed to estimate $\bar{F}_{i,1}(\cdot)$ by using the extended state, where $E_{i,1}$ represents the estimation error, $\phi_{i,1,1}$ is the estimation value for $e_{i,1}$, $\phi_{i,1,2}$ is the estimation for $\bar{F}_{i,1}(\cdot)$, $\sigma_{i,1,2} \in (0, 1)$, $\rho_{i,1,1}$ and $\rho_{i,1,2}$ are the positive constants to be chosen, and sign$(\cdot)$ is a signum function. The details about the ESO's convergence can be found in [10].

The first virtual control variable of the ith follower $\alpha_{i,2}$ is

$$\alpha_{i,2}(t) = -\frac{1}{b_{i,1}}\left[k_{i,0}w_i\xi_i + k_{i,1}e_{i,1}(t) + \phi_{i,1,2} + \dot{\delta}_i(t) - (1 - c_i)\dot{\hat{y}}_{di}\right], \tag{4.8}$$

with the adaptive law

$$\dot{\hat{y}}_{di} = -\gamma_i\xi_i - \mu_i\left(\hat{y}_{di} - y_{di}^0\right), \tag{4.9}$$

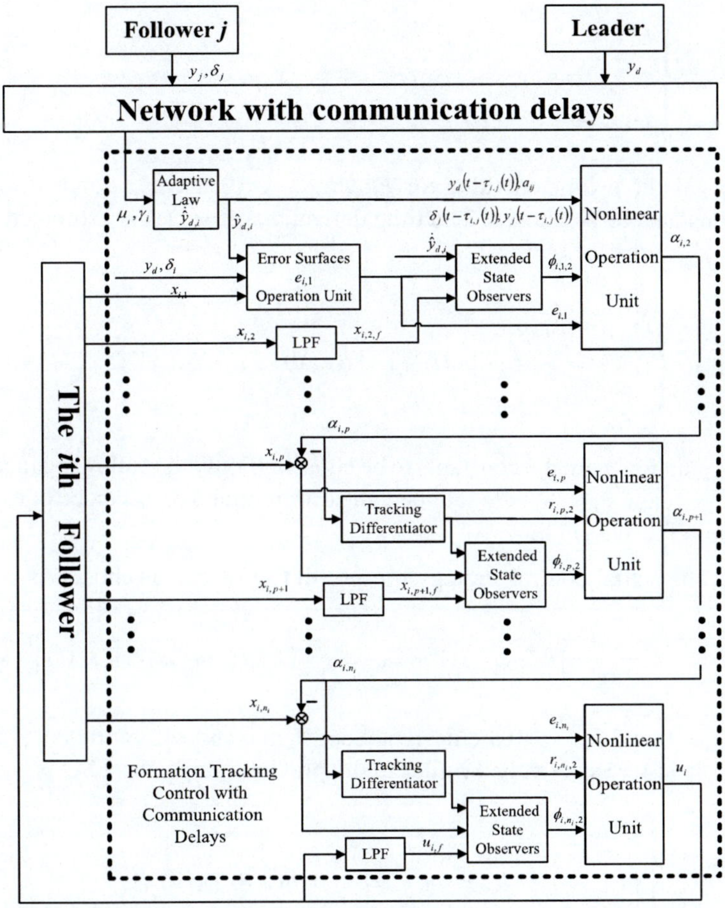

Fig. 4.2 The block diagram of the proposed formation tracking control with communication delays for the ith follower

where $\xi_i = \sum_{j \in \mathcal{N}_i} \left[y_i - \delta_i - y_j(t - \tau_{i,j}) + \delta_j \right] + c_i \left(y_i - \delta_i - y_d \right)$, $b_{i,1}$, $k_{i,0}$, w_i and $k_{i,1}$ are design positive constants, $\phi_{i,1,2}$ is the outputs of the ESO, and γ_i, μ_i and y_{di}^0 are positive constants.

Step i, p $(2 \le p \le n_i - 1)$: The virtual control variable of the ith follower $\alpha_{i,p+1}$ can be chosen as

$$\alpha_{i,p+1}(t) = -\frac{1}{b_{i,p}} \Big[k_{i,p} e_{i,p}(t) + \phi_{i,p,2} - r_{i,p,2} + b_{i,p-1} e_{i,p-1}(t) \Big], \tag{4.10}$$

where $b_{i,p} > 0$ and $k_{i,p} > 0$ are positive constants, $e_{i,p}(t) = x_{i,p} - \alpha_{i,p}(t)$, $\phi_{i,p,2}$ is the output signal of ESO (4.11) for the approximation of $\bar{F}_{i,p}$, and $r_{i,p,2}$ is the output signal of (4.12) for the estimation of the time derivative of $\alpha_{i,p}(t)$. In detail, the unknown function $\bar{F}_{i,p}$ is approximated by an ESO

$$\begin{cases} E_{i,p} = \phi_{i,p,1} - e_{i,p}, \\ \dot{\phi}_{i,p,1} = \phi_{i,p,2} - \rho_{i,p,1} E_{i,p} + b_{i,p} x_{i,p+1,f} - r_{i,p,2}, \\ \dot{\phi}_{i,p,2} = -\rho_{i,p,2} |E_{i,p}|^{\sigma_{i,p,2}} \mathrm{sign}(E_{i,p}), \end{cases} \tag{4.11}$$

where $\phi_{i,p,2}$ is the estimated value for $\bar{F}_{i,p}$, $\sigma_{i,p,2} \in (0, 1)$, $\rho_{i,p,1}$ and $\rho_{i,p,2}$ are the positive constants to be chosen. The time derivative of $\alpha_{i,p}(t)$ is estimated through the state $r_{i,p,2}$ in the following TD

$$\begin{cases} \dot{r}_{i,p,1} = r_{i,p,2}, \\ \dot{r}_{i,p,2} = -\pi_{i,p}^2 \mathrm{sign}(r_{i,p,1} - \alpha_{i,p}) |r_{i,p,1} - \alpha_{i,p}|^{\sigma_{i,p,1}} \\ \qquad - \pi_{i,p} r_{i,p,2}, \end{cases} \tag{4.12}$$

where $\sigma_{i,p,1}$, and $\pi_{i,p}$ are the constants to be chosen. If only the following inequalities $0 < \sigma_{i,p,1} < 1$, and $\pi_{i,p} > 0$ are satisfied, the differential of $\alpha_{i,p}$ can be of estimation by $r_{i,p,2}$ from the TD [11].

Step i, n_i: The actual control strategy for the ith follower u_i is chosen as

$$u_i(t) = -\frac{1}{b_{i,n_i}} \Big[k_{i,n_i} e_{i,n_i}(t) + \phi_{i,n_i,2} - r_{i,n_i,2} + b_{i,n_i-1} e_{i,n_i-1}(t) \Big], \tag{4.13}$$

where $b_{i,n_i} > 0$ and $k_{i,n_i} > 0$ are designed constants, and $\phi_{i,n_i,2}$ and $r_{i,n_i,2}$ are from (4.14) and (4.15), respectively. Similar to the previous steps, the ESO is

$$\begin{cases} E_{i,n_i} = \phi_{i,n_i,1} - e_{i,n_i}, \\ \dot{\phi}_{i,n_i,1} = \phi_{i,n_i,2} - \rho_{i,n_i,1} E_{i,n_i} + b_{i,n_i} u_{i,f} - r_{i,n_i,2}, \\ \dot{\phi}_{i,n_i,2} = -\rho_{i,n_i,2} |E_{i,n_i}|^{\sigma_{i,n_i,2}} \mathrm{sign}(E_{i,n_i}), \end{cases} \tag{4.14}$$

where $\rho_{i,n_i,1}$ and $\rho_{i,n_i,2}$ are design parameters, $\sigma_{i,n_i,2} \in (0, 1)$, and $\phi_{i,n_i,2}$ is the estimation value of the unknown function $\bar{F}_{i,n_i}$. The TD is designed as

$$\begin{cases} \dot{r}_{i,n_i,1} = r_{i,n_i,2}, \\ \dot{r}_{i,n_i,2} = -\pi_{i,n_i}^2 \mathrm{sign}(r_{i,n_i,1} - \alpha_{i,n_i}) |r_{i,n_i,1} - \alpha_{i,n_i}|^{\sigma_{i,n_i,1}} \\ \qquad - \pi_{i,n_i} r_{i,n_i,2}, \end{cases} \tag{4.15}$$

where $0 < \sigma_{i,n_i,1} < 1$ and $\pi_{i,n_i} > 0$.

From the above design procedure, we in a position to summarize the main result.

Theorem 4.1 *Consider the MAS consisting of one leader and uncertain non-affine nonlinear followers* (4.1) *with communication delays satisfying Assumptions 4.1 and 4.2. In the application of the control strategy* (4.13) *as well as the virtual controls* (4.8) *and* (4.10)*, all signals in the closed-loop system are ultimately bounded and the formation tracking error can be steered into a bounded residue around the origin with appropriate selection of parameters.*

Proof We construct a Lyapunov-Krasovskii function for the system

$$V(t) = \sum_{i=1}^{N} \Bigg[\sum_{p=1}^{n_i} \frac{1}{2} e_{i,p}^2 + \frac{k_{i,1} w_i (1 - c_i)}{2\gamma_i} \tilde{y}_{di}^2 + \frac{k_{i,0} w_i}{2} \sum_{j=1}^{N} \int_{t-\tau_{i,j}}^{t} (y_j(s) - \delta_j - y_d)^2 ds \Bigg], \tag{4.16}$$

where γ_i is a positive constants, $\tilde{y}_{di} = y_d - \hat{y}_{di}$. From (4.3), the virtual control variable (4.8), (4.10) and the actual control input of the ith follower (4.13), the time derivative of $V(t)$ yields

$$\begin{aligned} \dot{V}(t) = & \sum_{i=1}^{N} \Bigg\{ e_{i,1}(t) \Big[- k_{i,0} w_i \xi_i - k_{i,1} e_{i,1}(t) + (\bar{F}_{i,1} - \phi_{i,1,2}) \\ & + b_{i,1} l_{i,1}(t) \Big] + \sum_{p=2}^{n_i} e_{i,p}(t) \Big[- k_{i,p} e_{i,p}(t) + (\bar{F}_{i,p} - \phi_{i,p,2}) \\ & + (r_{i,p,2} - \dot{\alpha}_{i,p}) + b_{i,p} l_{i,p}(t) \Big] - \frac{k_{i,1} w_i (1 - c_i)}{\gamma_i} \tilde{y}_{di} \dot{\hat{y}}_{di} \\ & + \frac{k_{i,0} w_i}{2} \sum_{j=1}^{N} \Big[(y_j(t) - \delta_j - y_d)^2 - (y_j(t - \tau_{i,j}) - \delta_j - y_d)^2 \Big] \Bigg\}. \end{aligned} \tag{4.17}$$

By virtue of the adaptive law (4.9), $\tilde{y}_{di} \hat{y}_{di} \leq -\frac{1}{2} \tilde{y}_{di}^2 + \frac{1}{2} y_{di}^2$ and $b_{i,p} e_{i,p} l_{i,p}(t) \leq \frac{1}{2} e_{i,p}^2 + \frac{1}{2} b_{i,p}^2 l_{i,p,0}^2$, we can obtain

$$\begin{aligned} \dot{V}(t) \leq & \sum_{i=1}^{N} \Bigg\{ - (k_{i,1} - 1) e_{i,1}^2(t) - \sum_{p=2}^{n_i} \left(k_{i,p} - \frac{3}{2} \right) e_{i,p}^2(t) \\ & + \sum_{p=1}^{n_i} \frac{(\bar{F}_{i,p} - \phi_{i,p,2})^2}{2} + \sum_{p=2}^{n_i} \frac{(r_{i,p,2} - \dot{\alpha}_{i,p})^2}{2} \\ & - \frac{k_{i,1} w_i (1 - c_i) \mu_i}{2\gamma_i} \tilde{y}_{di}^2 + \frac{k_{i,1} w_i (1 - c_i) \mu_i}{2\gamma_i} \left(y_d - y_{di}^0 \right)^2 \\ & + \sum_{p=1}^{n_i} \frac{b_{i,p}^2 l_{i,p,0}^2}{2} - \frac{k_{i,0} w_i}{2} \sum_{j=1}^{N} [(y_i(t) - \delta_i) - (y_j(t - \tau_{i,j}) - \delta_j)]^2 \\ & - k_{i,0} c_i w_i (y_i - y_d - \delta_i)^2 - \frac{k_{i,0} w_i}{2} \\ & \sum_{j=1}^{N} \Big[(y_i(t) - \delta_i - y_d)^2 - (y_j(t) - \delta_j - y_d)^2 \Big] \Bigg\}, \end{aligned} \tag{4.18}$$

where μ_i is a constant.

From Ref. [10], we can obtain that the estimate error $\bar{F}_{i,p} - \phi_{i,p,2}$, $i = 1, \ldots, N$, $j = 1, \ldots, n_i$ in the ESOs can be sufficiently small by the appropriate selection of parameters $\sigma_{i,p,2}$, $\rho_{i,p,1}$ and $\rho_{i,p,2}$. Also, the estimate errors $r_{i,p,2} - \dot{\alpha}_{i,p}$ from the TDs can be sufficiently small with the appropriate parameters, $i = 1, \ldots, N$, $p = 2, \ldots, n_i$, and the details can be found in [11]. The error from the ESOs and TDs is denoted as $\varepsilon_1 = \sup\left(\sum_{i=1}^{N}\sum_{p=2}^{n_i}|r_{i,p,2} - \dot{\alpha}_{i,p}| + \sum_{i=1}^{N}\sum_{p=1}^{n_i}|\bar{F}_{i,p} - \phi_{i,p,2}|\right)$. The bound of the entire error is omitted for brevity and space limitation, and more details can be found in [1, 10, 12]. From Assumption 4.2 and Lemma 4.1, $w^T\mathcal{L} = 0$ holds. And, according to $-\frac{k_{i,0}w_i}{2}\sum_{i=1}^{N}\sum_{j=1}^{N}\left[(y_i(t)-\delta_i)-\left(y_j(t-\tau_{i,j})-\delta_j\right)\right]^2 - \sum_{i=1}^{N}k_{i,0}c_iw_i\,(y_i-y_d-\delta_i)^2 \le \left(\sum_{j=1}^{N}\int_{t-\tau_{i,j}}^{t}(y_j(s)-\delta_j-y_d)^2ds\right)^2$, the derivative of V yields

$$\dot{V} \le -V_e + S_\tau^2 + \varepsilon, \tag{4.19}$$

where $V_e = \sum_{i=1}^{N}[\left(k_{i,1}-1\right)e_{i,1}^2(t) + \sum_{p=2}^{n_i}(k_{i,p}-\frac{3}{2})e_{i,p}^2(t) + \frac{k_{i,1}w_i(1-c_i)\mu_i}{2\gamma_i}\tilde{y}_{di}^2]$, $\varepsilon = \frac{\varepsilon_1}{2} + \sum_{i=1}^{N}\frac{k_{i,1}w_i(1-c_i)\mu_i}{2\gamma_i}(y_d - y_{di}^0)^2 + \sum_{p=1}^{n_i}\frac{b_{i,p}^2 l_{i,p,0}^2}{2}$, $S_\tau = \sum_{j=1}^{N}\int_{t-\tau_{i,j}}^{t}(y_j(s) - \delta_j - y_d)^2ds$. From Definition 4.1, one has the system is ISpS. Select $\mathcal{K}$ class functions as $\kappa_1(s) = a_1s^2$, $\kappa_2(s) = a_2s^2$, $\kappa_3(s) = a_3s^2$, $\kappa_4(s) = s^2$, and V satisfies the defined conditions with $a_1 = \min\{\frac{1}{2}, \frac{k_{i,1}w_i\mu_i}{2\gamma_i}, \frac{k_{i,0}w_i}{2}\}$, $a_2 = \max\{\frac{1}{2}, \frac{k_{i,1}w_i\mu_i}{2\gamma_i}, \frac{k_{i,0}w_i}{2}\}$, $a_3 = \min\{k_{i,1}-1, k_{i,p}-\frac{3}{2}, \frac{k_{i,1}w_i\mu_i}{2\gamma_i}\}$. $L_\infty = \kappa_1^{-1}\circ\kappa_2\circ\kappa_3^{-1}\circ\kappa_4(s) = \sqrt{\frac{a_2}{a_1a_3}}s$. From $\dot{V} \le -V_e + S_\tau^2 + \varepsilon$, we know that there exist a $\mathcal{KL}$ class function $\alpha_v(\cdot)$ and a $\mathcal{K}$ class function $\beta_v(\cdot)$ such that $V_e \le \alpha_v(V_e(t_0), t - t_0) + \beta_v(S_\tau) + \varepsilon$, where $\beta_v(S_\tau) = \sqrt{\frac{a_2}{a_1a_3}}S_\tau$. The tracking error of the ith follower satisfies $|y_i - \delta_i - y_d| = |e_{i,1} - (1 - c_i)\tilde{y}_{di}| \le \sqrt{2e_{i,1}^2 + 2(1-c_i)\tilde{y}_{di}^2} \le \sqrt{4a_4V_e} \le \sqrt{4a_4(\alpha_v(V_e(t_0), t - t_0) + \beta_v(S_\tau) + \varepsilon)}$, where $a_4 = \max\{1, \frac{\gamma_i}{k_{i,1}w_i}\}$. Thus, formation tracking error is ultimately bounded.

Remark 4.1 From (4.3) and according to the design steps, if the virtual control variables $\alpha_{i,j+1}$ and control input u_i still involve in $x_{i,j+1}$ and u_i, $j = 1, \ldots, n_i - 1$, respectively, which will leads to an algebraic circle issue, and the control strategy is not feasible in practical applications. Inspired by the results in [8, 9], we introduce a Butterworth LPF into the ESO. $x_{i,j+1,f}$ and $u_{i,f}$ are used, instead of $x_{i,j+1}$ and u_i themselves, to approximate unknown dynamics, respectively. Then, $\alpha_{i,j+1}$ and u_i contain $x_{i,j+1,f}$ and $u_{i,f}$, respectively, and the algebraic circle behavior is ruled out.

Remark 4.2 The choice parameters of the ESO, the TD as well as the controllers should be chosen appropriately satisfying that $k_{i,1} > 1$, $k_{i,p} > \frac{3}{2}$, $\pi_{i,p} > 0$, $0 < \sigma_{i,p,1} < 1$, $0 < \sigma_{i,p,2} < 1$, $\rho_{i,p,1} > 0$, $\rho_{i,p,2} > 0$, $i = 1, \ldots, N$, $p = 1, \ldots, n_i$. And then, it is possible to adjust the error as small as desired.

Remark 4.3 In the situation that a_1, a_2, a_3 and a_4 remain constant, as $\tau_{i,j}$ increases, the interval $(t - \tau_{i,j}, t)$ lasts longer. Accordingly, the integral item $S_\tau = \sum_{j=1}^{N}\int_{t-\tau_{i,j}}^{t}(y_j(s) - \delta_j - y_d)^2ds$ is scaled up due to the fact that $(y_j(s) - \delta_j - y_d)^2 \ge 0$.

Thus, from $|y_i - \delta_i - y_d| \le \sqrt{4a_4(\alpha_v(V_e(t_0), t - t_0) + \beta_v(S_\tau) + \varepsilon)}$, the tracking error becomes larger. This indicates that the formation tracking performance becomes worse along with the increasement of $\tau_{i,j}$.

4.3 Simulation Examples

In order to illustrate the effectiveness of the formation tracking strategy, two simulation cases are presented in this section. The practical simulation is five mobile robots with the same dynamics, and the numerical one is four follower agents with different dynamics.

4.3.1 A Team of Mobile Robots

As shown the communication graph in Fig. 4.1, Node 1–Node 4 represent the followers and Node 0 is the leader. The dynamics of each robot are [13]

$$\begin{cases} M_i\ddot{x}_i + B_{x_i}\dot{x}_i = f_{x_i} - k_{xd_i}\dot{x}_i, \\ M_i\ddot{y}_i + B_{y_i}\dot{y}_i = f_{y_i} - k_{yd_i}\dot{y}_i, \end{cases} \tag{4.20}$$

where (x_i, y_i) is the coordinate of the ith robot, M_i is the mass of the ith robot, B_{x_i} and B_{y_i} are the damper coefficients of the ith mobile robot, f_{x_i} and f_{y_i} are the control forces of the ith mobile robot in two directions, k_{xd_i} and k_{yd_i} are coefficients. Furthermore, let $x_{i,1,1} = x_i, x_{i,1,2} = \dot{x}_i, u_{i,1} = f_{x_i}, x_{i,2,1} = y_i, x_{i,2,2} = \dot{y}_i, u_{i,2} = f_{y_i}$, and then, (4.20) is rewritten as

$$\begin{cases} \dot{x}_{i,1,1} = f_{i,1,1}(x_{i,1,1}, x_{i,1,2}), \\ \dot{x}_{i,1,2} = f_{i,1,2}(\overline{x}_{i,1,2}, u_{i,1}), \\ \dot{x}_{i,2,1} = f_{i,2,1}(x_{i,2,1}, x_{i,2,2}), \\ \dot{x}_{i,2,2} = f_{i,2,2}(\overline{x}_{i,2,2}, u_{i,2}), \\ y_{i,1} = x_{i,1,1}, \\ y_{i,2} = x_{i,2,1}, \end{cases} \tag{4.21}$$

where $(y_{i,1}, y_{i,2})$ is the coordinate of the ith follower, $\overline{x}_{i,1,2} = [x_{i,1,1}, x_{i,1,2}]^T, \overline{x}_{i,2,2} = [x_{i,2,1}, x_{i,2,2}]^T$, $u_{i,1}$ and $u_{i,2}$ are the control signals of the ith follower. $f_{i,1,1}(x_{i,1,1}, x_{i,1,2}) = x_{i,1,2}$, $f_{i,1,2}(\overline{x}_{i,1,2}, u_{i,1}) = -(B_{x_i} + k_{xd_i})x_{i,1,2}/M_i + u_{i,1}/M_i$, $f_{i,2,1}(x_{i,2,1}, x_{i,2,2}) = x_{i,2,2}$ and $f_{i,2,2}(\overline{x}_{i,2,2}, u_{i,2}) = -(B_{y_i} + k_{yd_i})x_{i,2,2}/M_i + u_{i,2}/M_i$ are unknown dynamics of the ith follower. In this example, $M_i = 1$, $B_{x_i} = 1$, $B_{y_i} = 1$, $k_{xd_i} = 9$, $k_{yd_i} = 9$. The initial values of Node 1–Node 4 are $[\overline{x}_{1,1,2}(0); \overline{x}_{1,2,2}(0)] = [5.5; 0; 2; 0]$, $[\overline{x}_{2,1,2}(0); \overline{x}_{2,2,2}(0)] = [5; 0; 1; 0]$, $[\overline{x}_{3,1,2}(0); \overline{x}_{3,2,2}(0)] = [4.5;$

Fig. 4.3 The outputs of the followers $y_{i,1}$, $y_{i,2}$, $i = 1, 2, 3, 4$, and the leader $y_{d,1}$, $y_{d,2}$ in Example A

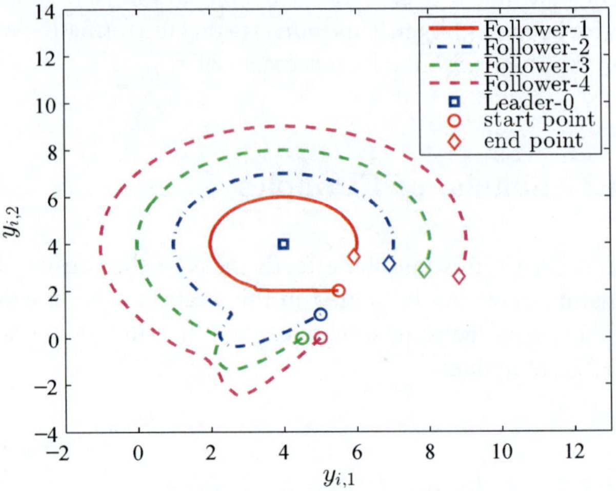

0; 0; 0], and $[\overline{x}_{4,1,2}(0); \overline{x}_{4,2,2}(0)] = [5; 0; 0; 0]$. The parameters in the control strategy are chosen as $\pi_{i,p,2} = 2$, $\sigma_{i,p,2,1} = 0.8$, $\rho_{i,p,1,1} = 10$, $\rho_{i,p,1,2} = 50$, $\sigma_{i,p,1,2} = 0.8$, $\rho_{i,p,2,1} = 10$, $\rho_{i,p,2,2} = 50$, $\sigma_{i,p,2,2} = 0.6$, $k_{i,p,0} = 0.1$, $w_{i,p} = 0.25$, $b_{i,p,1} = 1$, $b_{i,p,2} = 1$, $k_{i,p,1} = 5$, $k_{i,1,2} = 3$, $k_{i,2,2} = 5$, $\gamma_{i,p} = 0.01$, $\mu_{i,p} = 0.8$, $y^0_{di,p} = 4$, $i = 1, 2, 3, 4$, $j = 1, 2, 3, 4$, $p = 1, 2$. The initial values for ESOs and TDs are zero. The desired signals of the leaders are $y_{d,1} = 4$ and $y_{d,2} = 4$. The desired shape are $\delta_1 = [2\sin(t/5 - \pi); 2\cos(t/5 - \pi)]$, $\delta_2 = [3\sin(t/5 - \pi); 3\cos(t/5 - \pi)]$, $\delta_3 = [4\sin(t/5 - \pi); 4\cos(t/5 - \pi)) - 2.5]$ and $\delta_4 = [5\sin(t/5 - \pi); 5\cos(t/5 - \pi)]$. The communication delays are $\tau_{1,4} = 0.5$s, $\tau_{2,1} = 0.5$s, $\tau_{3,2} = 0.5$s and $\tau_{4,3} = 0.5$s.

Figure 4.3 shows the curves of the four followers and one leader i.e. the paths of mobile robots, and the outputs of the followers are able to track the output signal of the leader while building and keeping the desired shape via the control signals from the followers. The curves of the control input signals are shown in Fig. 4.4 The ESOs' estimation behaviors from the 1st follower are plotted in Fig. 4.5, and the uncertain dynamics of the follower can be estimated by the ESOs in a short period. Moreover, our proposed strategy is compared with the model-based strategy in Case 1 and Case 2 (Table 4.1, Example A), respectively. The tacking comparison is shown in Fig. 4.6. From Fig. 4.6, it is shown that the system performance governed by our control strategy is almost identical to model-based strategy in Case 1, and, as the system parameters are reduced by 20 percent in Case 2, the system behavior steered by the model-based strategy is worse than that by our strategy. It implies that, despite of uncertain parameters, the system with the control strategy in this chapter has better tracking performance than the model-based system has. This is due to the fact that unknown dynamics in the agents can be estimated and be compensated by ESOs effectively. The comparison of program execution time and system performance for the MASs are listed in Tables 4.2 and 4.3, where $\sum_{j=1}^{2} \bar{E}_{i,j}$ represents the total tracking error of the ith robot in two directions, and simulation time is is 50s. And the

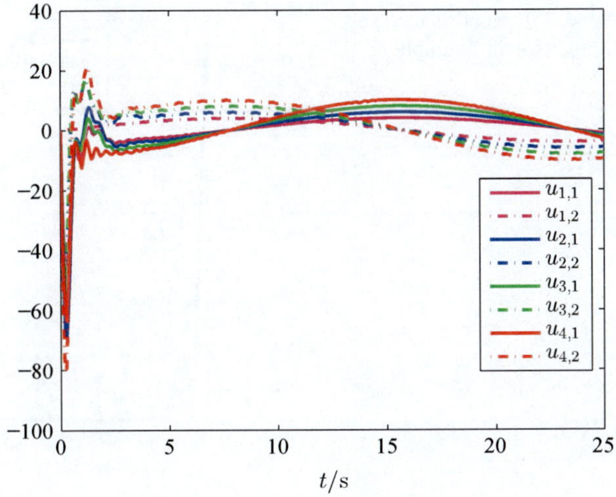

Fig. 4.4 The control signals in Example A

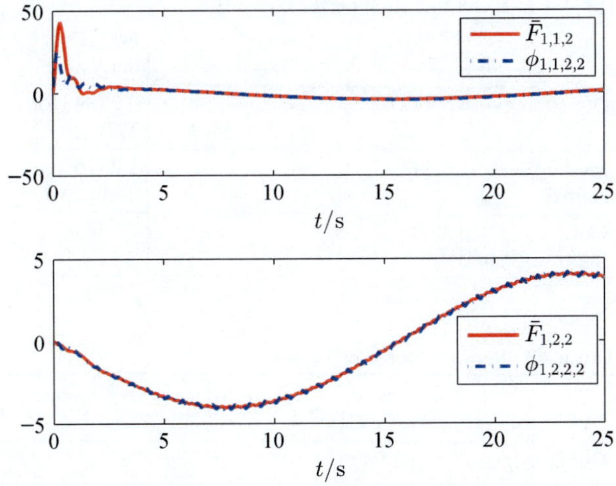

Fig. 4.5 The curves of $\bar{F}_{1,p,2}$, and $\phi_{1,p,2,2}$ in Example A, $p = 1, 2$

Table 4.1 Uncertain model parameters in Example A

Case	Uncertain model parameters
1	$100\%M_i$, $100\%B_{xi}$, $100\%B_{yi}$, $100\%k_{xd_i}$, $100\%k_{yd_i}$
2	$80\%M_i$, $80\%B_{xi}$, $80\%B_{yi}$, $80\%k_{xd_i}$, $80\%k_{yd_i}$

tests are carried out on a laptop with Intel(R) Core (TM) i5-4210M CPU @2.60GHz 2.60GHz.

Here, it is worth mentioning that the same condition here refers to that initial states and controller parameters, in the two control approaches, are the same, and only one difference, between the two control approaches, is the way how to get derivatives of virtual control variables. From Tables 4.2 and 4.3, the execution time of the proposed

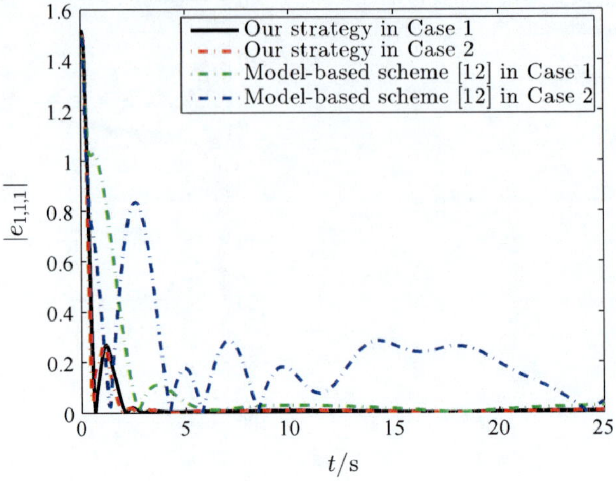

Fig. 4.6 Formation tracking comparison in Example A

Table 4.2 Execution time performance (s)

	Total	=	For TD or derivative	+	For others
Our strategy	5.4777	=	0.2784 (for TD)	+	5.1993
Model-based control	8.8429	=	1.3476(for derivative)	+	7.4953
Our improvement	38.06%		79.34%		–

Table 4.3 Performance comparison

	$\sum_{i=1}^{4}\sum_{j=1}^{2}\lvert u_{i,j}\rvert$	$\sum_{i=1}^{4}\sum_{j=1}^{2}\lvert\bar{E}_{i,j}\rvert$	$\sum_{i=1}^{4}\sum_{j=1}^{2}(\lvert\bar{E}_{i,j}\rvert^2+\lvert u_{i,j}\rvert^2)$
Our strategy	4.0435×10^4	339.2563	4.6256×10^5
Model-based control scheme	6.8390×10^4	387.8720	6.3086×10^5

strategy in this chapter is shorter than that of the scheme [5], and the two formation tracking performances are without significant difference. This is due to the fact that the TD is employed to reduce the complexity in the computation of the derivatives of complex nonlinear functions.

4.3.2 Numerical Example

The proposed time-varying formation tracking strategy is applied in a nonlinear MAS consisting of three followers and one leader, and the graph $\mathcal{G}_A$ of this example is shown in Fig. 4.7. The non-affine pure-feedback dynamics of the followers are

$$\begin{cases} \dot{x}_{i,1} = f_{i,1}(x_{i,1}, x_{i,2}), \\ \dot{x}_{i,2} = f_{i,2}(\overline{x}_{i,2}, u_i), \\ y_i = x_{i,1}, \end{cases} \tag{4.22}$$

where $f_{i,1}(x_{i,1}, x_{i,2}) = 2x_{i,2} + \sin(x_{i,1})x_{i,2}$, $f_{i,2}(\overline{x}_{i,2}, u_i) = x_{i,1}x_{i,2} + u_i + \left(2 - \sin(x_{i,1})\cos(x_{i,2})\right)u_i$, $i = 1, 2$, and $f_{j,1}(x_{j,1}, x_{j,2}) = x_{j,2} + x_{j,1}x_{j,2}$, $f_{j,2}(\overline{x}_{j,2}, u_j) = x_{j,1}x_{j,2} + (2 - \sin(t) - \exp(-t))\,u_j$, $j = 3, 4$, are unknown functions in the ith and jth follower. The initial values for the followers are $\overline{x}_{1,2}(0) = [5; 0]$, $\overline{x}_{2,2}(0) = [8; 0]$, $\overline{x}_{3,2}(0) = [3; 0]$, $\overline{x}_{4,2}(0) = [0; 0]$, and the desired signal of the leader is $y_d = 4$. The design parameters in the time-varying formation tracking controllers are chosen as $\pi_{i,2} = 2$, $\sigma_{i,2,1} = 0.4$, $\rho_{i,1,1} = 10$, $\rho_{i,1,2} = 50$, $\sigma_{i,1,2} = 0.8$, $\rho_{i,2,2} = 10$, $\rho_{i,2,2} = 50$, $\sigma_{i,2,2} = 0.8$, $\pi_{j,2} = 2$, $\sigma_{j,2,1} = 0.4$, $\rho_{j,1,1} = 10$, $\rho_{j,1,2} = 50$, $\sigma_{j,1,2} = 0.9$, $\rho_{j,2,2} = 10$, $\rho_{j,2,2} = 50$, $\sigma_{j,2,2} = 0.9$, $k_{i,0} = 0.1$, $k_{j,0} = 0.1$, $w_i = 0.25$, $w_j = 0.25$, $b_{i,1} = 1$, $b_{i,2} = 1$, $b_{j,1} = 1$, $b_{j,2} = 1$, $k_{1,1} = 5$, $k_{1,2} = 25$, $k_{2,1} = 2$, $k_{2,2} = 15$, $k_{j,1} = 5$, $k_{j,2} = 25$, $\gamma_i = 0.01$, $\gamma_j = 0.01$, $\mu_i = 0.8$, $\mu_j = 0.8$, $y_{di}^0 = 4$, $y_{dj}^0 = 4$, $i = 1, 2$, $j = 3, 4$. The initial values for ESOs and TDs are zero. The assigned outputs of three followers in the desired shape are $\delta_1 = 1 + 0.5\sin(t - \pi)$, $\delta_2 = 3 - \cos(t - \pi)$, $\delta_3 = -1 - 0.5\sin(t - \pi)$, and $\delta_4 = -3 + \cos(t - \pi)$. The communication delays are $\tau_{2,1} = 0.5$s, $\tau_{3,2} = 0.5$s, $\tau_{4,3} = 0.5$s and $\tau_{1,4} = 0.5$s.

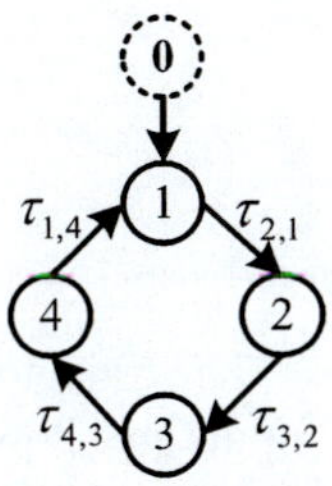

Fig. 4.7 The communication graph in Example B

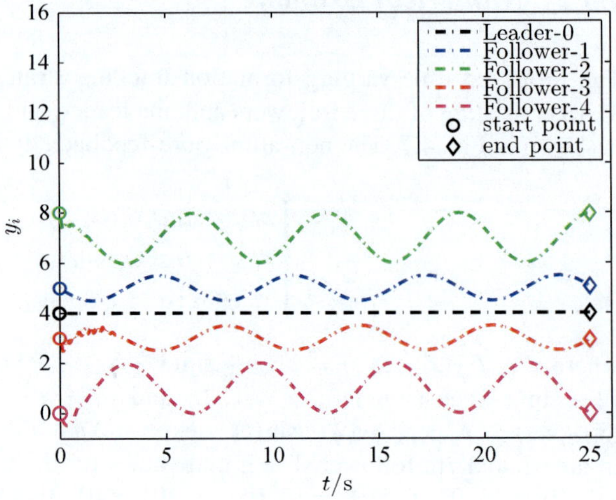

Fig. 4.8 The outputs of followers y_1, y_2, y_3, y_4 and leader y_d in Example B

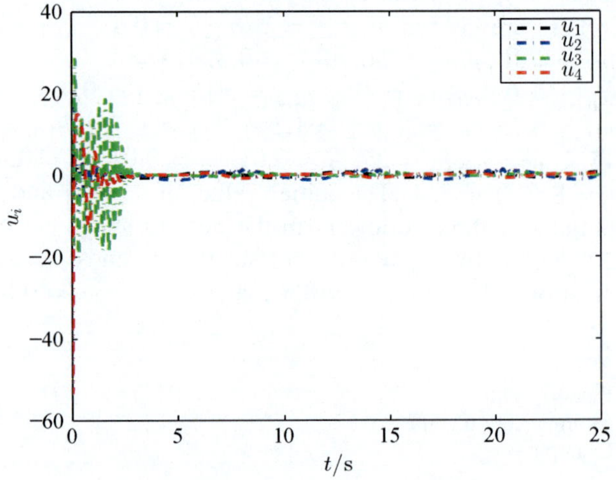

Fig. 4.9 The control input signals of Example B

The trajectories of the followers are plotted in Fig. 4.8, $i = 1, 2, 3, 4$. The control inputs are shown in Fig. 4.9. The ESOs' behaviors from the 1st follower with nominal and perturbation models are plotted in Fig. 4.10, and, from Fig. 4.10, it reveals that the ESO can compensate uncertain dynamics from perturbation model. The simulation results in Fig. 4.11 show the inverse relationship between the amplitude of $\tau_{2,1}$ and the formation tracking error of the 2nd follower, and this is consist with the analysis in Remark 4.3.

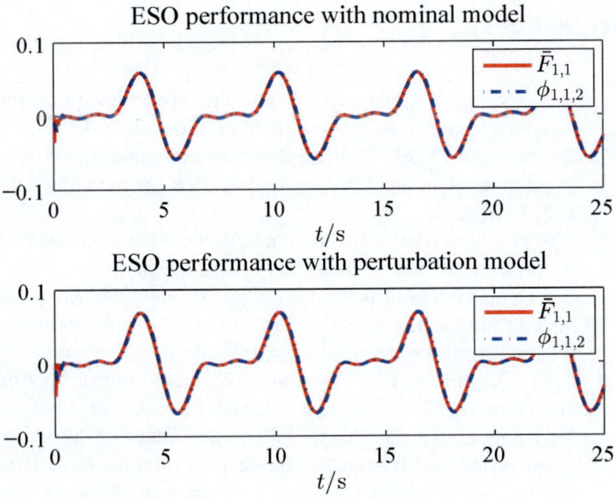

Fig. 4.10 The ESO performances with nominal and perturbation models in Example B

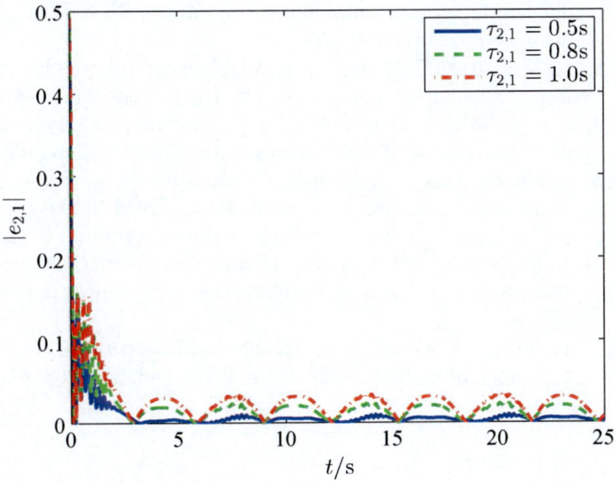

Fig. 4.11 The performance of $|e_{1,1}|$ with different delays in Example B

4.4 Conclusion

In this chapter, a time-varying formation tracking strategy has been presented for uncertain non-affine nonlinear MASs with communication delays. The ESO provides effective solutions for estimation the uncertainties in the MAS and for compensation for uncertain items in real time, and the TD is applied to estimate the derivative of complex nonlinear function. The proposed strategy is utilized to steer the output signal of followers to track the output signal of the leader in the desired formation shape.

References

1. C.H. Cheng, Y.A. Hu, J.H. Wu, Auto disturbance controller of non-affine nonlinear pure-feedback systems. Acta Automatica Sinica **40**(7), 1528–1536 (2014)
2. W. Wang, D. Wang, Z. Peng, Distributed containment control for uncertain nonlinear multi-agent systems in non-affine pure-feedback form under switching topologies. Neurocomputing **152**, 1–10 (2015)
3. P.V. Miroslav Krstic, I. Kanellakopoulos, *Nonlinear and Adaptive Control Design*. 1em plus 0.5em minus 0.4em (Wiley, New York, 1995)
4. G.C. Konstantopoulos, Q.-C. Zhong, B. Ren, M. Krstic, Bounded integral control of input-to-state practically stable nonlinear systems to guarantee closed-loop stability. IEEE Trans. Autom. Control **61**(12), 4196–4202 (2016)
5. L. He, X. Sun, Y. Lin, Distributed adaptive control for time-varying formation tracking of a class of networked nonlinear systems. Int. J. Control **90**(7), 1319–1326 (2017)
6. W. Wang, C. Wen, J. Huang, Distributed adaptive output feedback control of uncertain multi-agent systems with actuator faults and communication delays, in *Proceeding of 2016 Chinese Control and Decision Conference* (Yinchuan, 2016), pp. 7–13
7. W. Yu, G. Chen, M. Cao, J. Kurths, Second-order consensus for multiagent systems with directed topologies and nonlinear dynamics. IEEE Trans. Syst. Man Cybern. Part B (Cybernetics) **40**(3), 881–891 (2009)
8. A.-M. Zou, Z.-G. Hou, M. Tan, Adaptive control of a class of nonlinear pure-feedback systems using fuzzy backstepping approach. IEEE Trans. Fuzzy Syst. **16**(4), 886–897 (2008)
9. Y. Li, S. Tong, T. Li, Adaptive fuzzy output feedback dynamic surface control of interconnected nonlinear pure-feedback systems. IEEE Trans. Cybern. **45**(1), 138–149 (2014)
10. Y. Huang, J. Han, Analysis and design for the second order nonlinear continuous extended states observer. Chinese Sci. Bull. **45**(21), 1938 (2000)
11. H. J., Active disturbance rejection control technology: uncertainty estimation compensation control technology. 1em plus 0.5em minus 0.4em (National Defence Industry Press, 2008)
12. B.-Z. Guo, Z.-L. Zhao, On convergence of tracking differentiator. Int. J. Control **84**(4), 693–701 (2011)
13. H. Rezaee, F. Abdollahi, A decentralized cooperative control scheme with obstacle avoidance for a team of mobile robots. IEEE Trans. Ind. Electron. **61**(1), 347–354 (2013)

Chapter 5
Distributed Consensus Control of Nonlinear Multiagent Systems

Within the MASs framework, the relative states between neighbors can be acquired by some on-board sensors, and then the relative state saturations inevitably occur due to the limited sensing capabilities. This chapter investigates the consensus problem of nonlinear MASs subject to the relative state saturations. Utilizing the incidence matrix and the edge Laplacian, the consensus problem of nonlinear MASs with the relative state saturations can be cast into the stabilization problem of edge dynamics operating on the constrained set. Three types of protocols, namely continuous, intermittent and adaptive state feedback protocols are respectively proposed for achieving the constrained consensus, and meanwhile yielding consensus values. A consensus analysis is provided by virtue of state saturation theory, switched system theory, adaptive theory and Lyapunov stability theory. Output feedback protocol is also designed. Finally, the obtained results are applied to connectivity preservation for first order nonlinear MASs, despite the presence of limited communication range and input constraints. The theoretical results are validated by two simulation examples.

This chapter is organized as follows. Section 5.1 presents adaptive PI protocol for achieving consensus with relative state saturation constraints. The degenerated protocol for the special case and application in constrained preserving connectivity are provided in Sect. 5.2. In Sect. 5.3, the theoretical analyses are validated by a simulation example. Finally, Sect. 5.4 concludes the chapter.

5.1 Problem Formulation

Consider a group of agents consisting of agent, whose dynamics are described by

$$\dot{x}_i(t) = Ax_i(t) + Df(x_i(t), t) + Bu_i(t), i \in \mathbf{I}[1, N]. \tag{5.1}$$

where $x_i \in R^n$ is state, $u_i \in R^p$ is control input. Throughout this chapter, suppose that (A, B) is stabilizable. The continuously differentiable vector-valued function

D. Yue et al., *Distributed Cooperative Control and Communication for Multi-agent Systems*, https://doi.org/10.1007/978-981-33-6718-0_5

$f : R^n \times [0, +\infty) \rightarrow R^m$ represents the intrinsic nonlinear term, which is assumed to satisfy the following Lipschitz condition.

Assumption 5.1 There exists a nonnegative constant ρ , such that

$$\|f(a, t) - f(b, t)\| \leq \rho\|a - b\|, \forall a, b \in R^n, t \in R. \tag{5.2}$$

Assumption 5.2 The undirected graph G is connected.

Under Assumption 5.2, L has zero eigenvalue with multiplicity one, and all other eigenvalues are positive [1], and are ordered in increasing order as $0 \leq \lambda_2 \leq \cdots \leq \lambda_N$. Then, there exists an orthogonal matrix U such that $U^T LU = \Lambda = diag\{0, \lambda_2, \ldots, \lambda_N\}$, with $\lambda_i > 0$ for $i \in \mathbf{I}[2, N]$, and $U = [U_1, \ldots, U_N]$ with $U_1 = \mathbf{1}_N/\sqrt{N}$.

The edge Laplacian of an oriented graph G is defined as

$$\hat{L} = E^T E. \tag{5.3}$$

where the incidence matrix $E = [e_{ij}] \in R^{N\times M}$ is defined as

$$e_{ij} = \begin{cases} 1, & if\ v_i \text{ is the head of edge of } j \\ -1, & if\ v_i \text{ is the tail of edge of } j \\ 0, & \text{otherwise.} \end{cases}$$

The graph and edge Laplacian are real symmetric matrices, and the nonzero eigenvalues of $\hat{L}$ are equal to the non-zero eigenvalues of L.

In some practical MASs, the relative states $x_i - x_j$ can be obtained by some on-board sensors such as laser range finder. Then, the constraints on $x_i - x_j$ inevitably occur due to the limited sensing capabilities. In this chapter, assume that the relative state errors $x_i - x_j$ are confined into a closed hypercube in R^n. On the other hand, compared with a continuous protocol, an intermittent protocol and adaptive protocol are more popular, in that the former is practical, efficient and environmentally friendly, and the latter needs no any global information of the communication graph.

Motivated by these analyses, the goal of this chapter is to design three types of protocols, namely continuous, intermittent and adaptive protocols to achieve consensus in the sense of $\lim_{t\rightarrow\infty} \|x_i(t) - x_j(t)\| = 0$, under the constraint that the relative states $x_i - x_j$ operate on the closed hypercube in R^n. The key to deal with this type of constraints is translating the relative states into edge states, and further exploring edge dynamics operating on the closed hypercube, whose stability analysis and controller synthesis are solved by modifying the corresponding theory of a single system with state constraints [2, 3].

Before ending this section, the edge states and its hypercube are defined as follows. Let $[z_1^T, \ldots, z_M^T]^T = (E^T \otimes I_n)[x_1^T, \ldots, x_N^T]^T$. Together with definition of E, we know that the relative state $x_i - x_j$ can be represented by certain an edge state $z \in R^n$, i.e.,

$$z_m = \sum_{j=1}^{N} e_{jm} x_j, \, m \in \mathbf{I}[1, M], \tag{5.4}$$

where e_{jm} is the (j, m)th entry of the incidence matrix E. Assume that the constraints on $z_m \in R^n$ is component-wise, then its constrained set is defined by the following hypercube:

$$\Omega_m = \{z_m \in R^n | - \Delta_m \leq z_{k,m} \leq \Delta_m, m \in \mathbf{I}[1, M], k \in \mathbf{I}[1, n]\}. \tag{5.5}$$

where $z_{k,m}$ is k-th component of z_m, and the scalar $\Delta_m > 0$. The boundary of the set (5.5) is defined as $\partial\Omega_m = \bigcup_{k=1}^{n} \partial\Omega_{k,m}$, where $\partial\Omega_{k,m} = \{z_m \in R^n || z_{k,m}| = \Delta_m, |z_{l,m}| \leq \Delta_m, l \in \mathbf{I}[1, n] \backslash \{k\}\}$.

To achieve consensus of systems (5.1) with the constraints $z_m \in \Omega_m$ for $m \in \mathbf{I}[1, M]$, three types of protocols, namely continuous, intermittent and adaptive state feedback protocols is proposed in the following, and then the extension to output feedback case is also considered.

5.2 Continuous Consensus Protocol Design

Based on the relative state constraints, a continuous consensus protocol is proposed as

$$u_i(t) = \alpha K \sum_{j=1}^{N} a_{ij}(x_j - x_i). \tag{5.6}$$

where a_{ij} is the (i, j)th entry of the adjacency matrix A associated with the undirected graph G, $\alpha > 0$ is the gain constant, $K \in R^{m \times n}$ is the feedback gain matrix to be specified later. From the definitions (5.4), we get that $\sum_{j=1}^{N} a_{ij}(x_i - x_j) = \sum_{j=1}^{N} l_{ij} x_j = \sum_{j=1}^{N} \sum_{m=1}^{M} e_{im} e_{jm} x_j = \sum_{m=1}^{M} e_{im} z_m$. Then the protocol (5.6) with the relative states can be rewritten into one with the edge states:

$$u_i(t) = -\alpha K \sum_{m=1}^{M} e_{im} z_m. \tag{5.7}$$

Substituting (5.7) into (5.1) yields the following coupling node and edge dynamics, respectively:

$$\dot{x}_i(t) = A x_i(t) + D f(x_i(t), t) - \alpha B K \sum_{m=1}^{M} e_{im} z_m \tag{5.8}$$

and

$$\dot{z}_m = A z_m - \alpha B K \sum_{j=1}^{M} \hat{l}_{mj} z_j + D \sum_{j=1}^{N} e_{jm} f(x_j(t), t) \tag{5.9}$$

for $i \in \mathbf{I}[1, N]$, $m \in \mathbf{I}[1, M]$, where $\hat{L} = [\hat{l}_{mj}] \in R^{M \times M}$ is defined in (3). Considering the constraints (5.5), edge dynamics (5.9) can further expressed by

$$\begin{cases} \dot{z}_m = h(\xi_m), \\ \xi_m = Az_m - \alpha BK \sum_{j=1}^{M} \hat{l}_{mj} z_j + D \sum_{j=1}^{N} e_{jm} f(x_j(t), t). \end{cases} \tag{5.10}$$

where $z_m \in \Omega_m$ for $m \in \mathbf{I}[1, M]$, and the function $h(\cdot)$ represents the relative states constraints defined by $h(\xi_m) = [h(\xi_{1,m}), \ldots, h(\xi_{n,m})]^T$ with, for $k \in \mathbf{I}[1, n]$,

$$h(\xi_{k,m}) = \begin{cases} 0, \; if \; z_m \in \partial\Omega_{k,m}, \xi_{k,m} \cdot z_{k,m} > 0 \\ \xi_{k,m}, \text{ otherwise.} \end{cases} \tag{5.11}$$

The systems (5.8) and (5.10) can be recast in a compact form as:

$$\dot{x}(t) = (I_N \otimes A)x(t) + (I_N \otimes D)f(x(t), t) - \alpha(E \otimes BK)z, \tag{5.12}$$

and

$$\begin{cases} \dot{z} = h(\xi) \\ \xi = (I_M \otimes A - \alpha\hat{L} \otimes BK)z + (E^T \otimes D)f(x(t), t). \end{cases} \tag{5.13}$$

where $x = [x_1^T, \ldots, x_N^T]^T$, $z = [z_1^T, \ldots, z_N^T]^T$, $\xi = [\xi_1^T, \ldots, \xi_N^T]^T$, $f(x(t), t) = [f(x_1(t), t), \ldots, f(x_N(t), t)]^T$.

The function $h(\cdot)$ is called a state saturation [2, 3] that is used to constrain the state variables to a certain subspace. From its definition in (5.11), it is clear that saturation occurs in the edge state z_m, if $z_m \in \partial\Omega_{k,m}, \xi_{k,m} \cdot z_{k,m} > 0$. That is, when edge states $z_{k,m}$ are located on the boundary of the hypercube Ω_m , and tend to move outwards, edge states $z_{k,m}$ are penalized by the term $z_{k,m} = 0$. Otherwise, these two neighboring agents may go beyond the interaction rang and lose contact forever [4]. Hence, the saturation function $h(\cdot)$ forces the edge states to stay within its constrained set. To deal with the saturation $h(\cdot)$, the following lemma is given by modifying from [2, 3, 5].

Lemma 5.1 *Given (5.5), consider a symmetric positive definite matrix* $= P = [p_{ij}] \in R^{n \otimes n}$ *satisfying*

$$p_{ii} \geq \sum_{j=1, j \neq i}^{n} |p_{ij}|, i \in \boldsymbol{I}[1, N]. \tag{5.14}$$

Then, for any $z \in \Omega$ *and* $\xi \in R^{nM}$, *one has*

$$z^T(I_M \otimes P)(h(\xi) - \xi) \leq 0. \tag{5.15}$$

where $\Omega = \prod_{m=1}^{M} \Omega_m$ *is the Cartesian product.*

Proof To state more simply, we first give the index set representing the saturated components: $S = \{(i,m)|z_m \in \partial\Omega_{i,m}, \xi_{i,m} \cdot z_{i,m} > 0\}, for\ i \in \mathbf{I}[1,n], m \in \mathbf{I}[1,M]$, and its complement $S^C = \{(i,m)|(i,m) \notin S\}$. Then, from (5.11), we have $h(\xi_i,m) = \begin{cases} 0, if (i,m) \in S \\ \xi_{i,m}, if (i,m) \in S^C \end{cases}$. Then, it is not difficult to get that

$$
\begin{aligned}
& z^T (I_M \otimes P)(h(\xi) - \xi) \\
& = \sum\nolimits_{m=1}^{M} z_m^T P (h(\xi_m) - \xi_m) \\
& = \sum\nolimits_{m=1}^{M} \sum\nolimits_{j=1}^{n} \sum\nolimits_{i=1}^{n} z_{j,m} P_{ji} (h(\xi_{i,m}) - \xi_{i,m}) \\
& = \sum\nolimits_{m=1}^{M} \sum\nolimits_{i=1}^{n} \left(P_{ii} z_{i,m} + \sum\nolimits_{j=1,j\neq i}^{n} P_{ij} z_{j,m}\right) (h(\xi_{i,m}) - \xi_{i,m}) \\
& = \sum\nolimits_{(i,m)\in S} \left(P_{ii} z_{i,m} + \sum\nolimits_{j=1,j\neq i}^{n} P_{ij} z_{j,m}\right) (h(\xi_{i,m}) - \xi_{i,m}) \\
& \quad + \sum\nolimits_{(i,m)\in S^c} \left(P_{ii} z_{i,m} + \sum\nolimits_{j=1,j\neq i}^{n} P_{ij} z_{j,m}\right) (h(\xi_{i,m}) - \xi_{i,m}) \\
& = -\sum\nolimits_{(i,m)\in S} \left(P_{ij} z_{i,m} + \sum\nolimits_{j=1,j\neq i}^{n} P_{ij} z_{j,m}\right) \xi_{i,m} \\
& = -\sum\nolimits_{(i,m)\in S} \left(P_{ii} + \sum\nolimits_{j=1,j\neq i}^{n} P_{ij} \tfrac{z_{j,m}}{z_{i,m}}\right) z_{i,m} \cdot \xi_{i,m}
\end{aligned}
$$

Since for $(i,m) \in S$, $|z_{i,m}| = \Delta_m$, and $|z_{j,m}| \leq \Delta_m$, i.e., $\left|\frac{z_{j,m}}{z_{i,m}}\right| \leq 1$, i.e., , one has from (5.14) that

$$
P_{ii} + \sum_{j=1,j\neq i}^{n} P_{ij} \frac{z_{j,m}}{z_{i,m}} \geq P_{ii} - \sum_{j=1,j\neq i}^{n} |P_{ij}| \left|\frac{z_{j,m}}{z_{i,m}}\right| \geq P_{ii} - \sum_{j=1,j\neq i}^{n} |P_{ij}| \geq 0.
$$

On the other hand, for $(i,m) \in S$, $z_{i,m} \cdot \xi_{i,m} > 0$. Therefore, $\sum\nolimits_{(i,m)\in S} \left(P_{ii} + \sum\nolimits_{j=1,j\neq i}^{n} P_{ij} \frac{z_{j,m}}{z_{i,m}}\right) z_{i,m} \cdot \xi_{i,m} \geq 0$ holds, which implies that inequality (5.15) also holds.

The first main result is given now.

Theorem 5.1 *Under Assumptions 5.1 and 5.2, the continuous protocol (5.6) can realize consensus of nonlinear MASs (5.1) under the constraint (5.5) for all $z_m(0) \in \Omega_m$, $m \in \boldsymbol{I}[1,M]$, if there exists a scalar $\gamma > 0$ and a symmetric positive definite matrix $P = [p_{ij}] \in R^{n\times n}$, such that*

$$
p_{ii} \geq \sum_{j=1,j\neq i}^{n} |p_{ij}|, i \in \boldsymbol{I}[1,N]
$$

and

$$
PA + A^T P - 2PBB^T P + PDD^T P + \rho^2 I \leq -\gamma P. \tag{5.16}
$$

Choose $\alpha \geq 1/\lambda_2$, $K = B^T P$. Moreover, the consensus value is given by, for each $i \in \boldsymbol{I}[1,N]$, as $t \to \infty$,

$$x_i(t) \rightarrow \frac{1}{N}\sum_{i=1}^{N}\left[\exp(At)x_i(0) + \int_0^t \exp(A(t-s))Df(x_i(s), s)\,ds\right] \tag{5.17}$$

Proof The proof consists of two parts. The first part proves consensus in virtue of edge systems, while the second part seek the consensus value by resorting to node systems.

(Part 1) Consider a Lyapunov function $V = z^T(I_M \otimes P)z$. Note that $V \geq 0$, and $V = 0$ only when $z = 0$. Due to $z = (E^T \otimes I_n)x$, then $z = 0 \Leftrightarrow x_i - x_j = 0$ for $i, j \in \mathbf{I}[1, N]$, that is, the stability z of implies the consensus. The time derivative of V along (5.13) is given by

$$\begin{aligned}
\dot{V} &= 2z^T(I_M \otimes P)\dot{z} \\
&= 2z^T(I_M \otimes P)h(\xi) \\
&= 2z^T(I_M \otimes P)\xi + 2z^T(I_M \otimes P)(h(\xi) - \xi)
\end{aligned}$$

According to Lemma 1, we obtain that

$$\begin{aligned}
\dot{V} &\leq 2z^T(I_M \otimes P)\xi \\
&\leq \underbrace{z^T\left(I_M \otimes (PA + A^TP) - 2\alpha\hat{L} \otimes PBK\right)z}_{\Phi_1} + \underbrace{2z^T\left(E^T \otimes PD\right)f(x(t), t)}_{\Phi_2}
\end{aligned} \tag{5.18}$$

Let $\bar{x} = [\bar{x}_1^T, \ldots, \bar{x}_N^T]^T = (U^T \otimes I_n)x$, and together with the definitions (2) and (3), one has

$$\begin{aligned}
\Phi_1 &= x^T\left(EE^T \otimes (PA + A^TP) - 2\alpha E\hat{L}E^T \otimes PBB^TP\right)x \\
&= \bar{x}^T\left(U^TLU \otimes (PA + A^TP) - 2\alpha U^TL^2U \otimes PBB^TP\right)\bar{x} \\
&= \bar{x}^T\left(\Lambda \otimes (PA + A^TP) - 2\alpha\Lambda^2 \otimes PBB^TP\right)\bar{x} \\
&= \sum_{i=1}^{N}\bar{x}_i^T\lambda_i\left(PA + A^TP - 2\alpha\lambda_i PBB^TP\right)\bar{x}_i \\
&\leq \sum_{i=1}^{N}\bar{x}_i^T\lambda_i\left(PA + A^TP - 2PBB^TP\right)\bar{x}_i
\end{aligned}$$

where the last equality and inequality hold due to . Similar to the above derivation, we get that

$$
\begin{aligned}
\Phi_1 &\leq \sum_{i=1}^{N} \bar{x}_i^T \lambda_1 \left(PA + A^T P - 2PBB^T P\right) \bar{x}_i \\
&= \bar{x}^T \left(\Lambda \otimes \left(PA + A^T P - 2PBB^T P\right)\right) \bar{x} \\
&= \bar{x}^T \left(U^T L U \otimes \left(PA + A^T P - 2PBB^T P\right)\right) \bar{x} \\
&= x^T \left(L \otimes \left(PA + A^T P - 2PBB^T P\right)\right) x \\
&= z^T \left(I_M \otimes \left(PA + A^T P - 2PBB^T P\right)\right) z
\end{aligned}
\tag{5.19}
$$

Using the definition (1) and Lipschitz condition (5.2) gives

$$
\begin{aligned}
\Phi_2 &= 2z^T \left(I_M \otimes PD\right) \left(E^T \otimes I_n\right) f(x(t), t) \\
&\leq z^T \left(I_M \otimes PDD^T P\right) z + \left\| \left(E^T \otimes I_n\right) f(x(t), t) \right\|^2 \\
&\leq z^T \left(I_M \otimes PDD^T P\right) z + \rho^2 \left\| \left(E^T \otimes I_n\right) x \right\|^2 \\
&= z^T \left(I_M \otimes \left(PDD^T P + \rho^2 I\right)\right) z
\end{aligned}
\tag{5.20}
$$

where the second inequality holds due to the definition of E and Assumption 5.2. Inserting (5.19) and (5.20) into (5.18), and together with (5.16) yields $\hat{V} \leq -\gamma z^T (I_M \otimes P) z = -\gamma V$, which implies that all agents achieve consensus under the relative state constraints.

(Part 2) Next, we will derive the consensus value.

Letting $\bar{x} = \frac{1}{N} \sum_{i=1}^{N} x_i$, $\eta_i = x_i - \bar{x}$, and $\eta = [\eta_1^T, \ldots, \eta_N^T]^T$, we get $\eta = (H \otimes I_n) x$, where $H = I - \frac{\mathbf{1}\mathbf{1}^T}{N}$. It follows from (5.1) and (5.6) that

$$
\dot{\eta} = (I_N \otimes A - \alpha L \otimes BK)\eta + (H \otimes D) f(x(t), t). \tag{5.21}
$$

To stabilize (5.21), take a Lyapunov function $V = \eta^T (I_N \otimes P)\eta$, whose time derivative is given by

$$
\begin{aligned}
\dot{V} = &\underbrace{\eta^T \left(I_N \otimes \left(PA + A^T P - 2\alpha L \otimes PBB^T P\right)\right) \eta}_{\Phi_3} \\
&+ \underbrace{2\eta^T (H \otimes PD) f(x(t), t)}_{\Phi_4}
\end{aligned}
$$

Let $\bar{\eta} = (U^T \otimes I_n)\eta$, and one has $\bar{\eta}_1 = \sum_{i=1}^{N} \eta_i = 0$. Due to $U^T L U = \Lambda = diag\{0, \lambda^2, \ldots, \lambda_N\}$ and $\alpha\lambda_2 \geq 1$, then,

$$
\begin{aligned}
\Phi_3 &= \bar{\eta}^T \left(I_N \otimes PA + A^T P - 2\alpha \Lambda \otimes PBB^T P\right) \bar{\eta} \\
&\leq \bar{\eta}^T \left(I_N \otimes \left(PA + A^T P - 2PBB^T P\right)\right) \bar{\eta} \\
&= \eta^T \left(I_N \otimes \left(PA + A^T P - 2PBB^T P\right)\right) \eta
\end{aligned}
$$

and

$$\begin{aligned}\Phi_4 =&2\sum_{i=1}^{N}\eta_i^T PD\left(f\left(x_i(t),t\right)-\frac{1}{N}\sum_{j=1}^{N}f\left(x_j(t),t\right)\right)\\ =&2\sum_{i=1}^{N}\eta_i^T PD\left(f\left(x_i(t),t\right)-f(\bar{x},t)\right)\\ &+2\sum_{i=1}^{N}\eta_i^T PD\left(f(\bar{x},t)-\frac{1}{N}\sum_{j=1}^{N}f\left(x_j(t),t\right)\right)\\ \leq&\sum_{i=1}^{N}\left(\eta_i^T PDD^T P\eta_i+\|f\left(x_i(t),t\right)-f(\bar{x},t)\|^2\right)\\ \leq&\sum_{i=1}^{N}\left(\eta_i^T\left(PDD^T P+\rho^2 I\right)\eta_i\right)\end{aligned}$$

where the last inequality holds due to the Lipschitz condition (5.2). Thus, together with (5.16), one has $\dot{V}\leq-\gamma V$, which implies that $\eta_i\to 0\Leftrightarrow x_i\to\bar{x}$ as $t\to\infty$. From (5.8), one has $\dot{\bar{x}}_i=A\bar{x}_i+\frac{1}{N}\sum_{i=1}^{N}Df\left(x_i(t),t\right)-\alpha BK\frac{1}{N}\sum_{i=1}^{N}\sum_{m=1}^{M}e_{im}z_m$, The definition of E yields that $\mathbf{1}^T E=0$, i.e., $\sum_{i=1}^{N}\sum_{m=1}^{M}e_{im}z_m=0$. Then, $\dot{\bar{x}}_i=A\bar{x}_i+\frac{1}{N}\sum_{i=1}^{N}Df(x_i(t),t)$, solving it gives:

$$\bar{x}(t)=\exp(At)\bar{x}(0)+\frac{1}{N}\sum_{i=1}^{N}\int_0^t\exp(A(t-s))Df\left(x_i(s),s\right)ds \tag{5.22}$$

Submitting $\bar{x}(0)=\frac{1}{N}\sum_{i=1}^{N}x_i(0)$into (5.22) yields (5.17). This complete the proof.

Remark 5.1 With the help of the incidence matrix and the edge Laplacian, the consensus problem with the relative state saturations is converted into an edge agreement problem with the edge state saturations. Compared with the consensus problem, the edge agreement, rather than requiring the convergence to the agreement subspace, expects the edge dynamics to converge to the origin. Essentially, the evolution of edge dynamics (5.10) not only depends on the saturation function, but its own edge states and its neighbors edge states and the Lipschitz function. It can be observed from the proof that, under the same conditions, the edge agreement and average consensus are achieved simultaneously. Moreover, the final consensus values are given explicitly.

5.3 Intermittent Consensus Protocol Design

Intermittent protocols can capture topological variation and actuator failure [6]. In this subsection, we will design an intermittent consensus protocol under the constraint (5.5).

Under the relative state constraints, an intermittent consensus protocol is proposed as

$$u_i(t) = \begin{cases} \alpha K \sum_{j=1}^{N} a_{ij}\left(x_j - x_i\right), & t \in [l\omega, l\omega + \delta) \\ 0, & t \in [l\omega + \delta, (l+1)\omega) \end{cases} \tag{5.23}$$

where $l \in N$, scalars $\omega > \delta > 0$, other variables are the same in (5.6). Similar to the derivation in *Subsection A*, we obtain the following node and edge dynamics, respectively

$$\begin{cases} \dot{x}_i(t) = Ax_i(t) + Df\left(x_i(t), t\right) + \alpha BK \sum_{j=1}^{N} a_{ij}\left(x_j - x_i\right), t \in [l\omega, l\omega + \delta) \\ \dot{x}_i(t) = Ax_i(t) + Df\left(x_i(t), t\right), t \in [l\omega + \delta, (l+1)\omega) \end{cases} \tag{5.24}$$

and

$$\begin{cases} \dot{z}_m = h\left(\xi_m\right), \\ \xi_m = Az_m - \alpha BK \sum_{j=1}^{M} \hat{l}_{mj} z_j + D \sum_{j=1}^{N} e_{jm} f\left(x_j(t), t\right), t \in [l\omega, l\omega + \delta) \\ \dot{z}_m = h\left(\xi_m\right), \\ \xi_m = Az_m + D \sum_{j=1}^{N} e_{jm} f\left(x_j(t), t\right), t \in [l\omega + \delta, (l+1)\omega) \end{cases} \tag{5.25}$$

where $z_m \in \Omega_m$ for $m \in \mathbf{I}[1, M]$, $h(\cdot)$ is defined in (5.11), and other variables are the same in (5.8) and (5.10).

The consensus problem of systems (5.24) with the relative state constraints can be turned into the stabilization problem of edge dynamics (5.25) with the edge state saturations. Different form the continuous protocol (5.6), the intermittent protocol (5.23) cause the closed loop systems to be switched, and therefore the switched theory and the state saturation theory of a single system are pertinent to our present convergence analysis [7].

We are in position to give the second main result.

Theorem 5.2 *Under Assumptions 5.1 and 5.2, the intermittent protocol (5.23) can realize consensus of nonlinear multi-agent systems (5.24) under the constraint (5.5) for all $z_m(0) \in \Omega_m$, $m \in \boldsymbol{I}[1, M]$, if there exist positive scalars γ and β, and a symmetric positive definite matrices $P = [p_{ij}] \in R^{n \otimes n}$ and $Q = [q_{ij}] \in R^{n \otimes n}$, such that the following four inequalities hold:*

$$p_{ii} \geq \sum_{j=1, j \neq i}^{n} \left| p_{ij} \right|, q_{ii} \geq \sum_{j=1, j \neq i}^{n} \left| q_{ij} \right|, i \in \mathbf{I}[1, N], \tag{5.26}$$

$$PA + A^T P - 2PBB^T P + PDD^T P + \rho^2 I \leq -2\gamma P, \tag{5.27}$$

$$QA + A^T Q + QDD^T Q + \rho^2 I \leq 2BQ, \tag{5.28}$$

and

$$\delta/\omega > \beta/(\beta + \gamma) + \ln \mu/((\beta + \gamma)\omega) \tag{5.29}$$

where $\mu = max\{\lambda_{max}(P)/\lambda_{min}(Q), \lambda_{max}(Q)/\lambda_{min}(P)\}$. *Choose* $\alpha \geq 1/\lambda_2$, $K = B^T P$. *Moreover, the consensus value is the same in* (5.17).

Proof Construct the following multiple Lyapunov function for switched systems (5.25)

$$V = \begin{cases} \sum\limits_{m=1}^{M} z_m^T P z_m, \ t \in [l\omega, l\omega + \delta) \\ \sum\limits_{m=1}^{M} z_m^T Q z_m, \ t \in [l\omega + \delta, (l+1)\omega). \end{cases} \tag{5.30}$$

where positive definite matrices P and Q are the solutions of (5.26)–(5.28).

For $t \in [l\omega, l\omega + \delta), l \in N$, following the first part proof of Theorem 1, taking the time derivative of (5.30) along the trajectories of (5.25) gives,

$$\dot{V} \leq -2\gamma V \ for \ t \in [l\omega, l\omega + \delta). \tag{5.31}$$

where is define in (5.27).

For $t \in [l\omega + \delta, (l+1)\omega, l \in N$, similarly, together with Lemma 1, one has

$$\begin{aligned} \dot{V} &= \textstyle\sum_{m=1}^{M} 2z_m^T Q \dot{z}_m \\ &= \textstyle\sum_{m=1}^{M} 2z_m^T Q \xi_m + \sum_{m=1}^{M} 2z_m^T Q \left(h\left(\xi_m\right) - \xi_m\right) \\ &\leq \textstyle\sum_{m=1}^{M} 2z_m^T Q \xi_m \\ &= \textstyle\sum_{m=1}^{M} z_m^T \left(QA + A^T Q\right) z_m + \sum_{m=1}^{M} \sum_{j=1}^{N} 2e_{jm} z_m^T QDf\left(x_j(t), t\right). \end{aligned} \tag{5.32}$$

On the other hand, noticing the derivation of (5.20), we have

$$\begin{aligned} &\textstyle\sum_{m=1}^{M} \sum_{j=1}^{N} 2e_{jm} z_m^T QDf\left(x_j(t), t\right) \\ &= 2z^T \left(E^T \otimes QD\right) f(x(t), t) \\ &\leq \textstyle\sum_{m=1}^{M} z_m^T \left(QDD^T Q + \rho^2 I\right) z_m. \end{aligned} \tag{5.33}$$

Substituting (5.33) into (5.32), and noting (5.28) yields

$$\dot{V} \leq 2\beta V, \ for \ t \in [l\omega + \delta, (l+1)\omega). \tag{5.34}$$

Note that the switched systems (5.25) switches at $t = l\omega$ and $t = l\omega + \delta, l \in N$. Then, we obtain that

$$\begin{aligned} V(\omega) &\le \mu \exp(2\beta(\omega - \delta))V(\delta) \\ &\le \mu^2 \exp(2\beta(\omega - \delta) - 2\gamma\delta)V(0) \\ &= \exp(-\tau)V(0) \end{aligned}$$

where $\tau = 2\gamma\delta - 2\beta(\omega - \delta) - 2\ln\mu$. From (5.29), one has $\tau > 0$. By recursion, for any $l \in N$, one gets $V(l\omega) \le exp(-l\tau)V(0)$. For any $t > 0$, there exists $r \in N$ an such that $r\omega \le t \le (r+1)\omega$. Then one has

$$\begin{aligned} V(t) &\le \exp(2\beta\omega)V(r\omega) \\ &\le \exp(2\beta\omega)\exp(-r\tau)V(0) \\ &\le \exp(2\beta\omega)V(0)\exp\left(-\frac{\tau t}{\omega + 1}\right), \end{aligned}$$

which implies that under the constraint (5.5), the states of agents exponentially converge to the same values. Following the second part proof of Theorem 1, the consensus values can be also analytically derived. The proof completes.

5.4 Adaptive Consensus Protocol Design

In (5.6) and (5.23), the fixed coupling weight depends on the global spectra information, please see $\alpha \ge 1/\lambda_2$. To remove this condition, an adaptive protocol for adjusting the coupling weights is proposed as follows [8]:

$$\begin{aligned} u_i(t) &= \alpha_i(t)K\textstyle\sum_{j=1}^{N} a_{ij}\left(x_j - x_i\right) \\ \dot{\alpha}_i(t) &= v_i\left(\textstyle\sum_{j=1}^{N} a_{ij}\left(x_j - x_i\right)\right)^T \Gamma\left(\textstyle\sum_{j=1}^{N} a_{ij}\left(x_j - x_i\right)\right) \end{aligned} \tag{5.35}$$

where $V_i > 0$ denote the scaling factor to adjust the change rate of $\alpha_i(t)$, K and Γ are the feedback gain matrices to be determined later. We obtain the following edge dynamics

$$\begin{cases} \dot{z} = h(\xi), \\ \xi = \left(I_M \otimes A - E^T\alpha(t)E \otimes BK\right)z + \left(E^T \otimes D\right)f(x(t), t), \end{cases}$$

where $\alpha(t) = diag\{\alpha_1(t), \ldots, \alpha_N(t)\}$, all other variables are the same in (5.13).

Theorem 5.3 *Under Assumptions 5.1 and 5.2, the adaptive protocol (5.35) can realize consensus of nonlinear MASs (5.1) under the constraint (5.5) for all* $z_m(0) \in \Omega_m$, $m \in \boldsymbol{I}[1, M]$, *if there exist a symmetric positive definite matrix* $P = [p_{ij}] \in R^{n \otimes n}$, *such that*

$$p_{ii} \geq \sum_{j=1, j \neq i}^{n} \left| p_{ij} \right|, \ i \in \boldsymbol{I}[1, N],$$

and

$$PA + A^T P - 2PBB^T P + PDD^T P + \rho^2 I < 0. \tag{5.36}$$

Take $K = B^T P$ *and* $\Gamma = PBB^T P$.

Proof Construct a Lyapunov function

$$V = z^T (I_M \otimes P) z + \sum_{i=1}^{N} (\alpha_i(t) - \alpha_0)^2 / v_i$$

where α_0 is a positive constant with $\alpha_0 \geq 1/\lambda_2$. one has

$$\dot{V} \leq z^T \left(I_M \otimes \left(PA + A^T P \right) - 2E^T \alpha(t) E \otimes PBB^T P \right) z + \Phi_2 + \Phi_5$$

where Φ_2 is defined in (5.18), and

$$\begin{aligned}
\Phi_5 &= 2\sum_{i=1}^{N} (\alpha_i(t) - \alpha_0) \dot{\alpha}_i(t)/v_i \\
&= 2\sum_{i=1}^{N} (\alpha_i(t) - \alpha_0) \left(\sum_{j=1}^{N} a_{ij} \left(x_j - x_i \right) \right)^T PBB^T P \left(\sum_{j=1}^{N} a_{ij} \left(x_j - x_i \right) \right) \\
&= 2 (L \otimes I_n x)^T \left((\alpha(t) - \alpha_0 I_N) \otimes PBB^T P \right) (L \otimes I_n x) \\
&= 2z^T \left(E^T (\alpha(t) - \alpha_0 I_N) E \otimes PBB^T P \right) z
\end{aligned} \tag{5.37}$$

Thus, $\dot{V} \leq z^T \left(I_M \otimes \left(PA + A^T P \right) - 2\alpha_0 \hat{L} \otimes PBB^T P \right) z + \Phi_2$, the rest proofs are similar to that of Theorem 1, and thus is omitted here for brevity.

5.5 Extension to Output Feedback Case

In some applications, the relative state between the neighbors are not available, and then each agent need to estimate its own state by using its output. In such a circumstance, the relative state estimation can be acquired by sensors, and the constraints in the relative state estimation inevitably arise.

Each agents output is given by $y_i = Cx_i$, where $y_i \in R^q$.

Assumption 5.3 The triple (A, B, C) is controllable and observable.

An observer-based protocol is proposed

$$\begin{cases} \dot{\hat{x}}_i(t) = A\hat{x}_i(t) + Df\left(\hat{x}_i(t), t\right) + Bu_i(t) - F\left(y_i - C\hat{x}_i(t)\right), \\ u_i(t) = \alpha K \sum_{j=1}^{N} a_{ij}\left(\hat{x}_j - \hat{x}_i\right), \quad i \in \mathbf{I}[1, N], \end{cases} \tag{5.38}$$

where $\hat{x}_i$ is the estimate of x_i . Let $\tilde{x}_i = x_i - \hat{x}_i$, and $\hat{z} = (E^T \otimes I^n)\hat{x}$, where $\hat{z} = [\hat{z}_1^T, \ldots, \hat{z}_M^T]^T$ and $\hat{x} = [\hat{x}_1^T, \ldots, \hat{x}_N^T]^T$. It follows from (5.1) and (5.38) that

$$\dot{\tilde{x}}_i = (A + FC)\tilde{x}_i + Df\left(x_i(t), t\right) - Df\left(\hat{x}_i(t), t\right) \tag{5.39}$$

and

$$\begin{cases} \dot{\hat{z}} = h(\tilde{\xi}) \\ \tilde{\xi} = \left(I_M \otimes A - \alpha\hat{L} \otimes BK\right)\hat{z} + \left(E^T \otimes D\right) f(\hat{x}(t), t) - \left(E^T \otimes FC\right)\tilde{x} \end{cases} \tag{5.40}$$

where $\hat{z}_m \in \Omega_m$ for $m \in \mathbf{I}[1, M]$, and the saturation function $h(\cdot)$ is defined as

$$h\left(\hat{\xi}_{k,m}\right) = \begin{cases} 0, \text{ if } \hat{z}_m \in \partial\Omega_{k,m}, \hat{\xi}_{k,m} \cdot \hat{z}_{k,m} > 0, \\ \hat{\xi}_{k,m}, \text{ otherwise}. \end{cases}$$

Theorem 5.4 *Under Assumptions 5.1, 5.2 and 5.3, the observer-based protocol (39) can realize consensus of nonlinear MASs (5.1) under $\hat{z}_m \in \Omega_m$ for all $\hat{z}_m(0) \in \Omega_m, m \in \boldsymbol{I}[1, M]$, if there exist a scalar $\gamma > 0$ and a symmetric positive definite matrix $P = [p_{ij} \in R^{n \otimes n}]$ and Q , such that*

$$p_{ii} \geq \sum_{j=1, j\neq i}^{n} \left|p_{ij}\right|, i \in \boldsymbol{I}[1, N],$$

$$PA + A^T P - 2PBB^T P + PDD^T P + \rho^2 I \leq -\gamma P \tag{5.41}$$

and

$$QA + A^T Q - 2C^T C + QDD^T Q + \rho^2 I \leq -Q.$$

Take $\alpha \geq 1/\lambda_2$, $K = B^T P$ and $F = -Q^{-1}C^T$.

Proof Construct a Lyapunov function

$$V = \varepsilon\hat{z}^T\left(I_M \otimes P\right)\hat{z} + \tilde{x}^T\left(I_N \otimes Q\right)\tilde{x},$$

where ε is a positive constant to be determined later. Following the proof of Theorem 1 yields

$$\begin{aligned}
&2\hat{z}^T (I_M \otimes P)\dot{\hat{z}} \\
&\leq \hat{z}^T \left(I_M \otimes (PA + A^T P) - 2\alpha \hat{L} \otimes PBB^T P\right)\hat{z} \\
&\quad +2\hat{z}^T (E^T \otimes PD) f(\hat{x}(t), t) - 2\hat{z}^T (E^T \otimes PFC)\tilde{x} \\
&\leq \hat{z}^T (I_M \otimes (PA + A^T P - 2PBB^T P + PDD^T P + \rho^2 I))\hat{z} \\
&\quad -2\hat{z}^T (E^T \otimes PFC)\tilde{x} \\
&\leq -\gamma \hat{z}^T (I_M \otimes P)\hat{z} - 2\hat{z}^T (E^T \otimes PFC)\tilde{x},
\end{aligned}$$

and

$$\begin{aligned}
&2\sum_{i=1}^N \tilde{x}_i^T Q\dot{\tilde{x}}_i \\
&= \sum_{i=1}^N \begin{pmatrix} \tilde{x}_i^T \left(Q(A + FC) + (A + FC)^T Q\right)\tilde{x}_i \\ +2\tilde{x}_i^T QD \left(f(x_i(t), t) - f(\hat{x}_i(t), t)\right) \end{pmatrix} \\
&= \sum_{i=1}^N \begin{pmatrix} \tilde{x}_i^T \left(QA + A^T Q - 2C^T C\right)\tilde{x}_i + \tilde{x}_i^T QDD^T Q\tilde{x}_i^T \\ + \left\| f(x_i(t), t) - f(\hat{x}_i(t), t)\right\|^2 \end{pmatrix} \\
&= \sum_{i=1}^N \tilde{x}_i^T \left(QA + A^T Q - 2C^T C + QDD^T Q + \rho^2 I\right)\tilde{x}_i \\
&\leq -\sum_{i=1}^N \tilde{x}_i^T Q\tilde{x}_i
\end{aligned}$$

Combining these two parts, one has

$$\begin{aligned}
\dot{V} &\leq \begin{pmatrix} \hat{z} \\ \tilde{x} \end{pmatrix}^T \begin{pmatrix} -\gamma\varepsilon I_M \otimes P & -\varepsilon E^T \otimes PFC \\ -\varepsilon E \otimes C^T F^T P & -I_N \otimes Q \end{pmatrix} \begin{pmatrix} \hat{z} \\ \tilde{x} \end{pmatrix} \\
&\triangleq \begin{pmatrix} \hat{z} \\ \tilde{x} \end{pmatrix}^T \Psi \begin{pmatrix} \hat{z} \\ \tilde{x} \end{pmatrix}.
\end{aligned}$$

Choose a sufficient small positive constant ε , such that $-I_N \otimes Q + \varepsilon^2 E \otimes C^T F^T P (\gamma\varepsilon I_M \otimes P)^{-1} E^T \otimes PFC < 0$, and one has from Schur Lemma that $\Psi < 0$, i.e.,$\dot{V} < 0$, for $\hat{z} \neq 0$ and $\tilde{x} \neq 0$. This implies that as $t \to \infty$, one has $x_i - \hat{x}_i \to 0, x_j - \hat{x}_j \to 0$ *and* $\hat{x}_i - \hat{x}_j \to 0$, i.e., $x_i - x_j \to 0$. The proof is completed.

Remark 5.2 Different from [9–11] , in which the effects of relative state constraints are not considered, this chapter investigates consensus problem under the relative state saturations. By contrast, the additional condition need be imposed here that the positive definite solution matrices of (5.16), (5.27), (5.28) and (5.41) are diagonally dominant. Diagonal positive definite solution matrices are needed if we extend from the component-wise constraint to norm bounded constraints, which can be observed from the proof of Lemma 1.

5.6 Applications in Constrained Preserving Connectivity

In this subsection, the obtained results are applied to one-order nonlinear MASs, we solve an average consensus problem while preserving the connectedness, under limited communication range and input constraints.

Systems are given by $\dot{x}_i = f(x_i, t) + u_i, i \in \mathbf{I}[1, N]$, where $x_i \in R^n$, and $f(x_i, t)$ satisfying (5.2).

In practice, the communication between two agents is determined by their distance. The distance between agent i and j connected by an edge m is expressed by

$$\|z_m\| = \sqrt{\left(x_i - x_j\right)^T \left(x_i - x_j\right)} \geq 0.$$

Suppose that the communication range is $\Delta > 0$, then agent i can communicate with j if and only if $z_m \in \Omega_m$, where

$$\Omega_m \triangleq \{z_m \ \|z_m\| \leq \Delta, m \in \mathbf{I}[1, M]\}. \tag{5.42}$$

We further assume that each agents input satisfies: $\|u_i\| \leq \bar{u}$. A protocol is proposed as

$$u_i = -\alpha \sum_{m=1}^{M} e_{im} z_m \tag{5.43}$$

and one has

$$\begin{cases} \dot{z}_m = h\left(\xi_m\right), \\ \xi_m = \sum_{j=1}^{N} e_{jm} f\left(x_j(t), t\right) - \alpha \sum_{j=1}^{M} \hat{l}_{mj} z_j. \end{cases}$$

where all the variables are the same in (5.10), except that $h(\xi_m)$ is redefined as

$$h\left(\xi_m\right) = \begin{cases} 0, \ \text{if } z_m \in \partial\Omega_m, z_m^T \cdot \xi_m > 0 \\ \xi_m, \ \text{otherwise} \end{cases} \tag{5.44}$$

The following Lemma show that the connectedness can be preserved under the proposed protocol (5.43).

Lemma 5.2 *Consider the protocol* (5.43) *with the communication constraints* (5.42). *If, for all* $m \in I[1, M]$, $z_m(0) \in \Omega_m$, *then* $z_m \in \Omega_m$ *for* $t \geq 0$. *That is, the constrained set is* Ω_m *invariant for all* $z_m, m \in I[1, M]$.

Proof In fact, two connected agents can remain connected if for $t \geq 0$, $\frac{d\|z_m\|}{dt} \leq 0$ when $\|z_m\| = \Delta$. Then we consider

$$\frac{d\left\|z_m\right\|}{dt} = \frac{z_m^T \dot{z}_m}{\|z_m\|} = \frac{z_m^T h\left(\xi_m\right)}{\|z_m\|} = \begin{cases} 0, \ \text{if } z_m \in \partial\Omega_m, z_m^T \cdot \xi_m > 0 \\ \frac{z_m^T \xi_m}{\|z_m\|}, \ \text{if } z_m \in \partial\Omega_m, z_m^T \cdot \xi_m \leq 0 \end{cases}$$

where the last equality holds due to (5.44). This complete proof.

Corollary 5.1 *Under Assumptions* 5.1 *and* 5.2. *For all* $z_m(0) \in \Omega_m, m \in I[1, M]$ *, under the protocol* (5.43), *the average consensus is reached while preserving the connectedness if* $\alpha > (1 + \rho^2)/(2\lambda_2)$. *Moreover, the consensus value is given by* $\frac{1}{N} \sum_{i=1}^{N} \left(x_i(0) + \int_0^t f\left(x_i(\tau), \tau\right) \mathrm{d}\tau\right)$.

Proof Choose $V = \sum_{m=1}^{M} z_m^T z_m$, and then

$$\begin{aligned}
\dot{V} &= 2\sum_{m=1}^{M} z_m^T \dot{z}_m = 2\sum_{m=1}^{M} z_m^T h\left(\xi_m\right) \overset{(1)}{\leq} 2\sum_{m=1}^{M} z_m^T \xi_m \\
&= 2z^T\left(\left(E^T \otimes I_n\right) f\left(x,t\right) - \alpha\left(E^T E \otimes I_n\right) z\right) \\
&\overset{(2)}{\leq} \left(1+\rho^2\right) z^T z - 2\alpha z^T\left(E^T E \otimes I_n\right) z \\
&= \left(1+\rho^2\right) x^T\left(L \otimes I_n\right) x - 2\alpha x^T\left(L^2 \otimes I_n\right) x \\
&\leq \sum_{i=2}^{N}\left(1+\rho^2 - 2\alpha\lambda_2\right)\lambda_i \bar{x}_i^T \bar{x}_i
\end{aligned}$$

where $\overset{(1)}{\leq}$ and $\overset{(2)}{\leq}$ respectively hold due to Lemma 1 and (5.2). Then, $\dot{V} \leq 0$, and $\dot{V} = 0 \Leftrightarrow \bar{x}_i = 0, i \in \mathbf{I}[2, N]$. By LaSalle invariance principle, one has $\bar{x}_i = 0$ as $t \to \infty$, for $\mathbf{I}[2, N]$. since $U_1 = \mathbf{1}_N/\sqrt{N}$, we have $\lim_{t\to\infty} x(t) = \lim_{t\to\infty}\left(U \otimes I_n\right)\bar{x}(t) = \mathbf{1}_N \bar{x}_1(t)/\sqrt{N}$, where $\dot{\bar{x}}_1 = (1/\sqrt{N})\sum_{i=1}^{N} f\left(x_i, t\right)$.

We next further construct the coupling weights α such that the input constraints are also guaranteed. Since the constrained set Ω_m is invariance for all $z_m, m \in I[1, M]$, i.e., $\|z_m\| \leq \Delta$. Thus, $\|u_i\| \leq \alpha \sum_{m=1}^{M} \|e_{im}\|\|z_m\| \leq \alpha\Delta\bar{d} \leq \bar{u}$, where $\bar{d} = \max\limits_{i\in I[1,N]}\{\sum_{m=1}^{M}\|e_{im}\|\}$ denotes the maximal number of neighbors of all agents. Take α from the set

$$\{\alpha \mid \underline{\alpha} < \alpha \leq \bar{\alpha}\}\ with\ \underline{\alpha} = \left(1+\rho^2\right)/\left(2\lambda_2\right), \bar{\alpha} = \bar{u}/(\Delta\bar{d}), \tag{5.45}$$

then, average consensus is reached while preserving the connectedness and input constraints.

Remark 5.3 For given communication rage, mobile agents may create or lost their connectivity at any moment, and then the network topology is time-varying. Thus, in (5.45), the gain parameter α need to adjust dynamically according to variables $\underline{\alpha}$ and $\bar{\alpha}$. On the other hand, compared with network structure , it is easy to verify topological connectivity by the second small eigenvalue of Laplacian.

5.7 Simulation Examples

In this section, two examples of applications are provided to illustrate the effectiveness of Theorem 5.3 and Corollary 5.1 respectively.

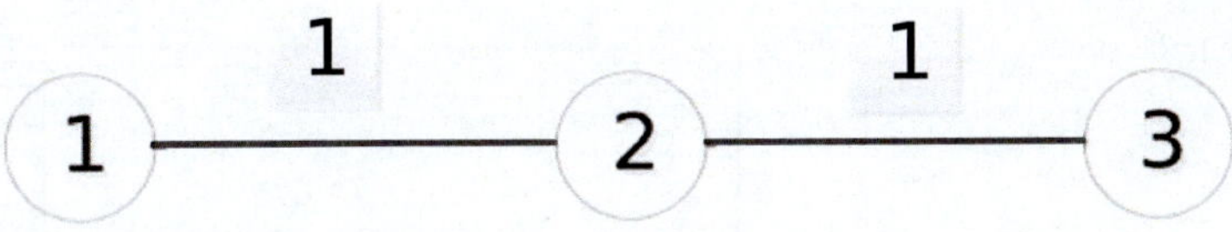

Fig. 5.1 An undirected network

5.7.1 Application to Aircraft Team

The communication topology is shown in Fig. 5.1. Without loss of generality, we assume its incidence matrix as

$$E = \begin{pmatrix} 1 & 0 \\ -1 & 1 \\ 0 & -1 \end{pmatrix}$$

In [12–14], the dynamical equation of the F-18 aircraft can be given as the decoupled linearized form:

$$\begin{bmatrix} \dot{x}_{1,i} \\ \dot{x}_{2,i} \end{bmatrix} = \begin{bmatrix} -1.175 & 0.9871 \\ -8.458 & -0.8776 \end{bmatrix} \begin{bmatrix} x_{1,i} \\ x_{2,i} \end{bmatrix} + \begin{bmatrix} 0 \\ 0.33 \sin x_{2,i} \end{bmatrix} + \begin{bmatrix} -0.194 & -0.03593 \\ -19.29 & -3.803 \end{bmatrix} \begin{bmatrix} u_{1,i} \\ u_{2,i} \end{bmatrix}$$

where $x_{1,i}$ and $x_{2,i}$ denotes angle of attack and pitch rate, respectively, and $u_{1,i}$, $u_{2,i}$ are the control inputs. The nonlinear term $0.33 \sin(x_{2,i})$ is Lipschitz, and $\rho = 0.33$.

In this condition, solving matrix inequality (5.36) yields

$$P = \begin{bmatrix} 0.1133 & 0.0076 \\ 0.0076 & 0.0191 \end{bmatrix}, K = \begin{bmatrix} -0.1692 & -0.3700 \\ -0.0331 & -0.0729 \end{bmatrix}$$

$$\Gamma = \begin{bmatrix} 0.0297 & 0.0650 \\ 0.0650 & 0.1422 \end{bmatrix}$$

The initial states of agents are assumed as

$$\begin{bmatrix} x_{1,1}(0) & x_{1,2}(0) & x_{1,3}(0) \\ x_{2,1}(0) & x_{2,2}(0) & x_{2,3}(0) \end{bmatrix} = \begin{bmatrix} -0.1 & 0.3 & -0.2 \\ 0.3 & -0.2 & 0.1 \end{bmatrix}$$

$$\begin{bmatrix} \alpha_1(0) & \alpha_2(0) & \alpha_3(0) \end{bmatrix} = \begin{bmatrix} 0.4 & -0.3 & 0.5 \end{bmatrix}$$

Take $\Delta_1 = \Delta_2 = 0.5$ and $\upsilon_1 = \upsilon_2 = \upsilon_3 = 3$. From Figs. 5.2 and 5.3, attitude synchronization is achieved despite presenting relative state saturation. Moreover, under the effect of this constraint, edge states are confined on the square $[-0.5, 0.5] \times$

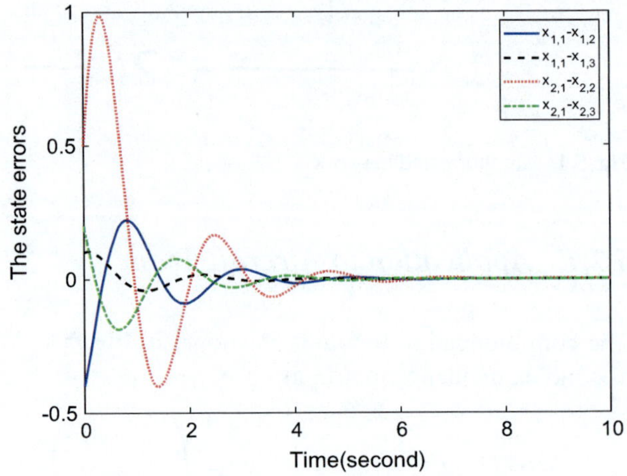

Fig. 5.2 The node state errors

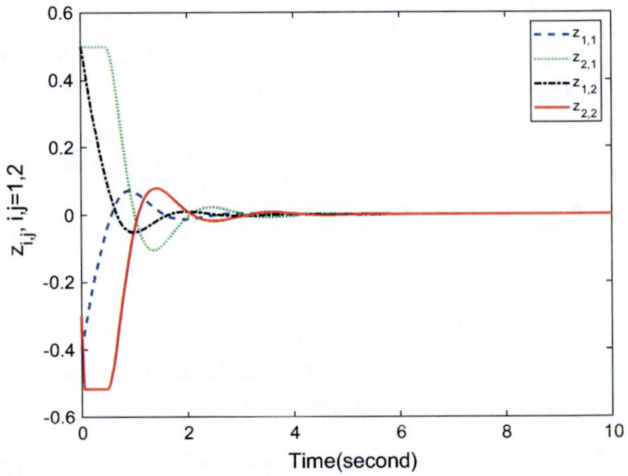

Fig. 5.3 The edge states

$[-0.5, 0.5]$. Once this constraint is removed, some edge state runs out of this square, which can be observed from Figs. 5.4 and 5.5. Finally, the trajectories of adaptive gains are attached in Fig. 5.6.

5.7.2 *Application to Connectivity Preservation*

Consider two-dimensional systems:

$$\begin{bmatrix} x_{1,i} \\ x_{2,i} \end{bmatrix} = \begin{bmatrix} 0 \\ 0.33 \sin x_{2,i} \end{bmatrix} + \begin{bmatrix} u_{1,i} \\ u_{2,i} \end{bmatrix}, i \in \mathbf{I}[1, 9]$$

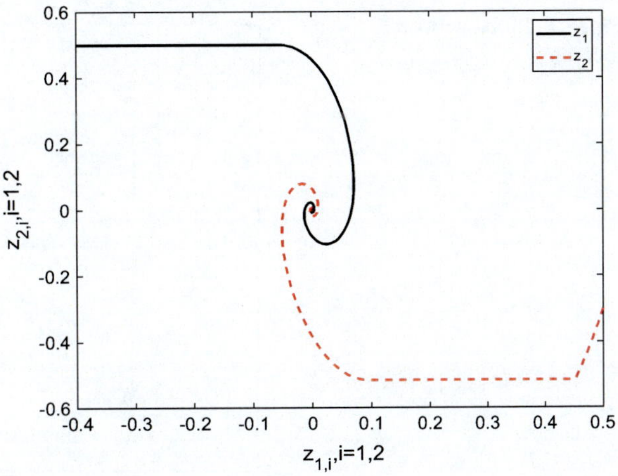

Fig. 5.4 The phase diagram of edge states with saturations

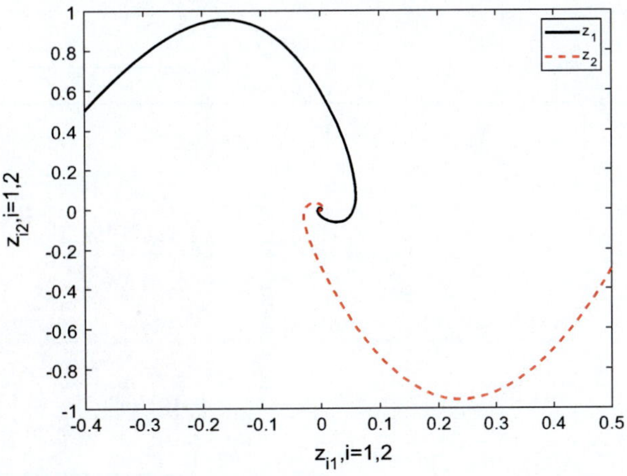

Fig. 5.5 The phase diagram of edge states with saturations

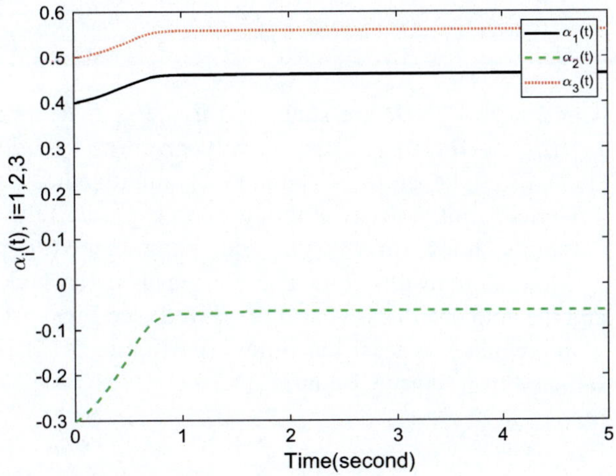

Fig. 5.6 The adaptive gain

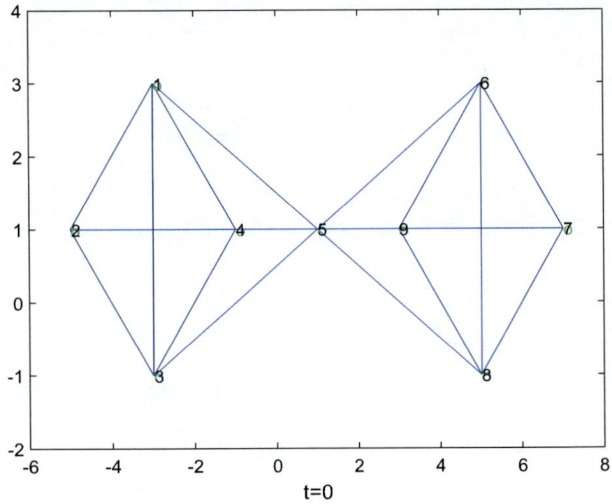

Fig. 5.7 Initial network topology

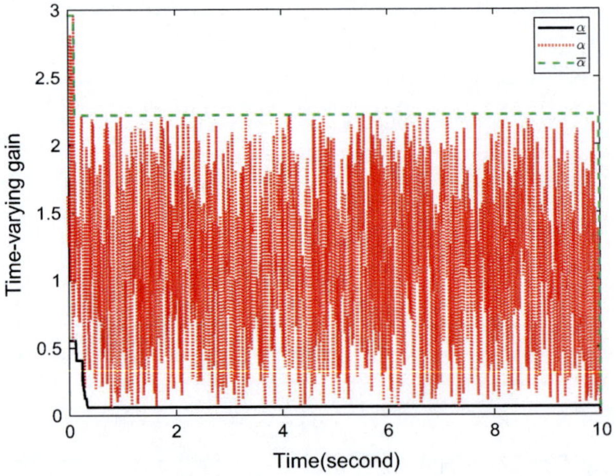

Fig. 5.8 Time-varying gain parameter

where $x_{1,i}, x_{2,i} \in R$ are state, and $u_{1,i}, u_{2,i} \in R$ are inputs. Initial states, $[x_{1,i}(0), x_{2,i}(0)]$, $i \in \mathbf{I}[1, 9]$ are respectively given as $[-3, 3]$, $[-5, 1]$, $[-3, -1]$, $[-1, 1]$, $[1, 1]$, $[5, 3]$, $[7, 1]$, $[5, -1]$ and [3,1], and together with communication range $\Delta = 4.5$, yields initial network topology in Fig. 5.7. Take $\bar{u} = 80$. According to (5.45), randomly choose time-varying gain parameter $\alpha(t)$, plotted in Fig. 5.8.

Simulation results show that consensus is reached (see Figs. 5.9, 5.10 and 5.11), and the connectedness is still preserved (see Figs. 5.12 and 5.13), although the network switches at least ten times (see Figure 5.13). The input constraints are also satisfied from Figs. 5.14 and 5.15.

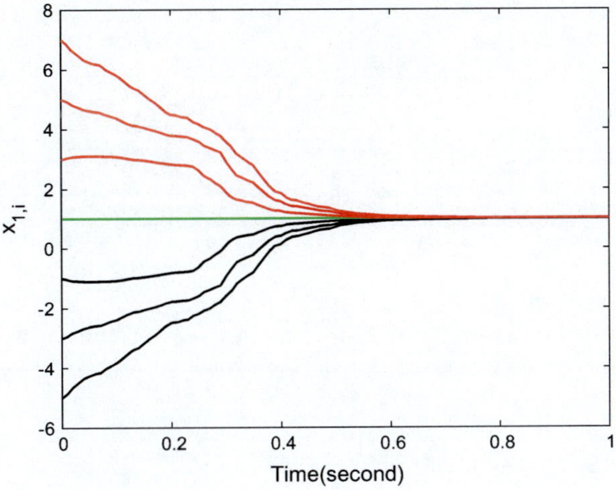

Fig. 5.9 The first component states

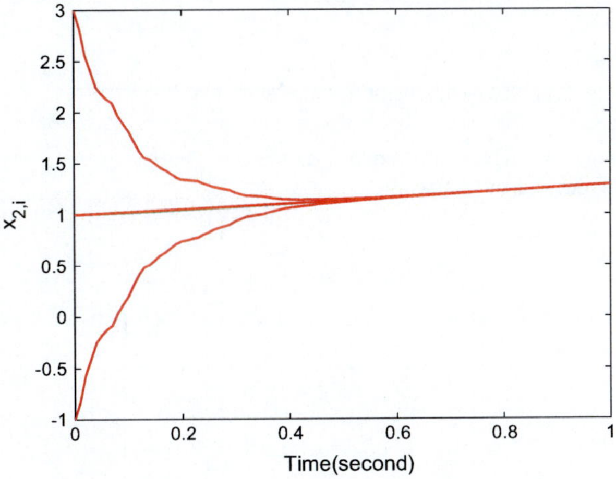

Fig. 5.10 The second component states

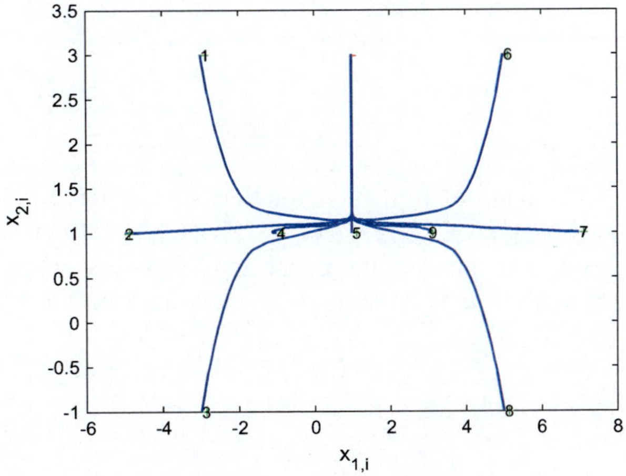

Fig. 5.11 Two-dimensional states (+ and − denote starting and terminal points, respectively)

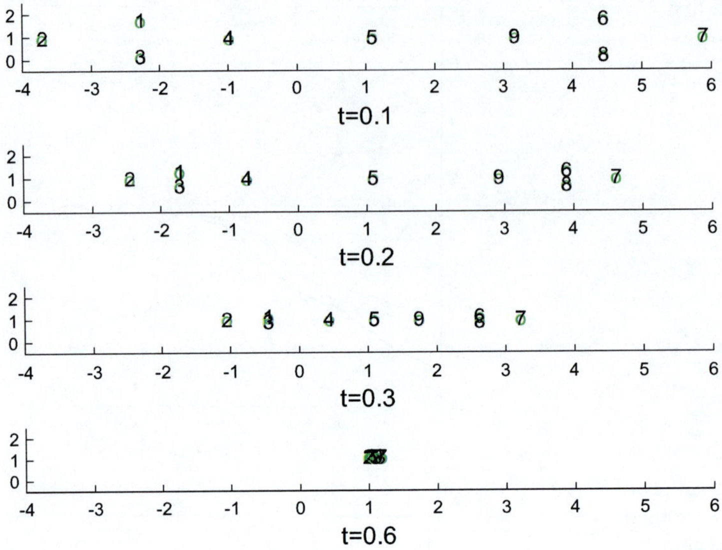

Fig. 5.12 Two-dimensional states at certain times

Fig. 5.13 The second small eigenvalue of Laplacian

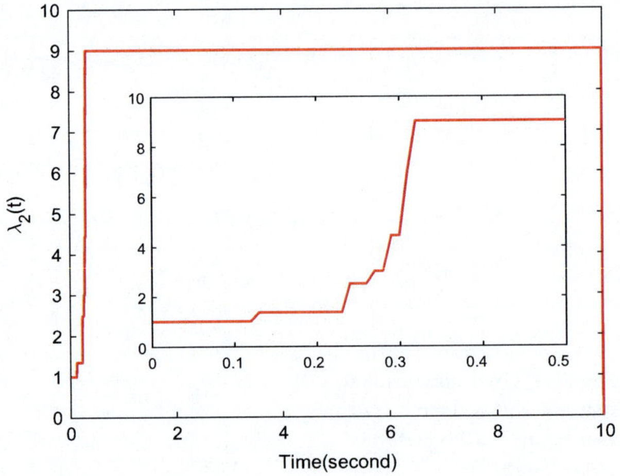

In addition, further simulation shows that for initial states $[-5, 3]$, $[-7, 1]$, $[-5, -1]$, $[-3, 1]$, $[1, 1]$, $[7, 3]$, $[9, 1]$, $[7, -1]$ and $[5,1]$, even though the initial graph is connected, the connectedness is lost when the standard consensus algorithm $u_i(t) = \sum_{j=1}^{N} \alpha_{ij}(x_j - x_i)$ is used, which can be seen from Fig. 5.16.

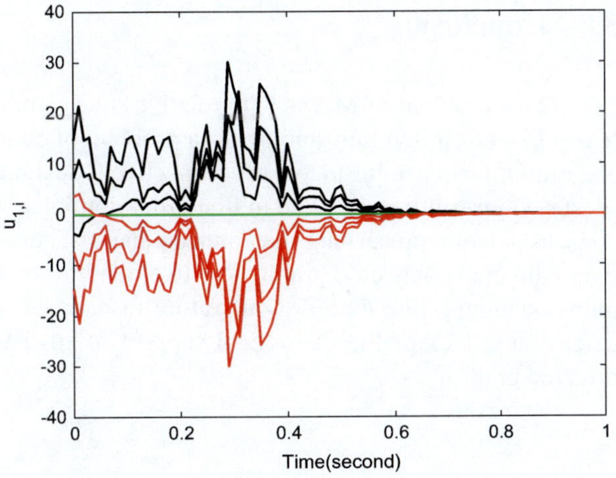

Fig. 5.14 The first component inputs

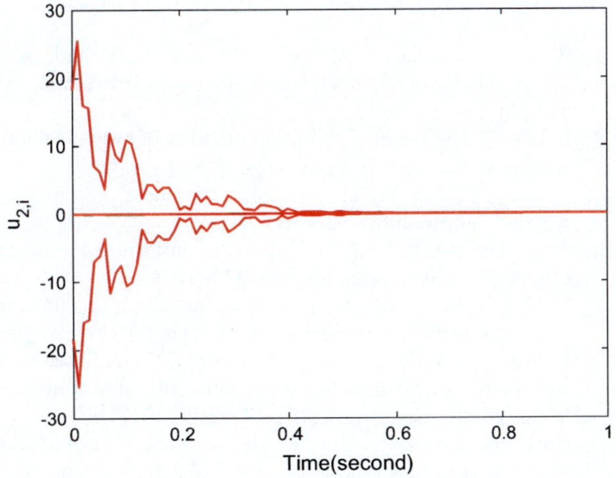

Fig. 5.15 he second component inputs

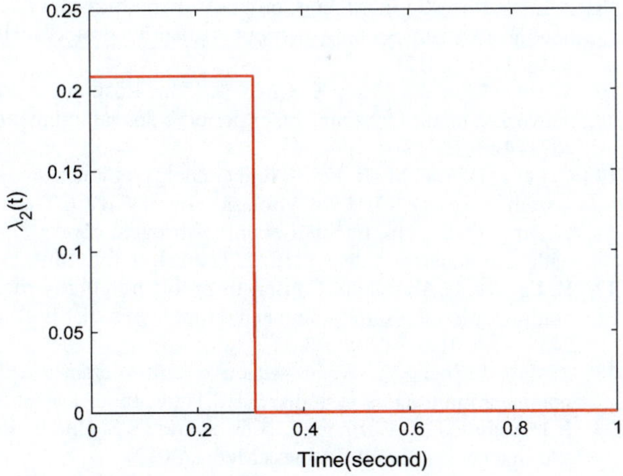

Fig. 5.16 The second small eigenvalue of Laplacian, under the standard consensus algorithm

5.8 Conclusion

Consensus problem of MASs with relative state saturations was studied. This problem is first converted into stabilization problem of constrained edge dynamics. State and output feedback protocols are respectively designed for achieving average consensus. Our results are applied to first order nonlinear MASs, and also solve average consensus while preserving the connectedness. These developments were based on the undirected connected graph, how to extend to the directed case, specially to the united spanning tree case are our future researches. The technical obstacle of the extension is lack of the theoretical support of edge Laplacian associated with the directed graphs.

References

1. R. Agha, M. Rehan, C.K. Ahn, G. Mustafa, S. Ahmad, Adaptive distributed consensus control of one-sided lipschitz nonlinear multiagents. IEEE Trans. Syst. Man Cybern. Syst. **49**(3), 568–578 (2017)
2. L. Hou, A.N. Michel, Asymptotic stability of systems with saturation constraints. IEEE Trans. Autom. Control **43**(8), 1148–1154 (1998)
3. H. Fang, Z. Lin, Stability analysis for linear systems under state constraints. IEEE Trans. Autom. Control **49**(6), 950–955 (2004)
4. X. Ai, S. Song, K. You, Second-order consensus of multi-agent systems under limited interaction ranges. Automatica **68**, 329–333 (2016)
5. Y.-H. Lim, H.-S. Ahn, Consensus under saturation constraints in interconnection states. IEEE Trans. Autom. Control **60**(11), 3053–3058 (2015)
6. B. Wang, J. Wang, B. Zhang, W. Chen, Z. Zhang, Leader-follower consensus of multivehicle wirelessly networked uncertain systems subject to nonlinear dynamics and actuator fault. IEEE Trans. Autom. Sci. Eng. **15**(2), 492–505 (2017)
7. S. Hu, D. Yue, Q.-L. Han, X. Xie, X. Chen, C. Dou, Observer-based event-triggered control for networked linear systems subject to denial-of-service attacks. IEEE Trans. Cybern. **50**(5), 1952–1964 (2019)
8. Z. Li, W. Ren, X. Liu, M. Fu, Consensus of multi-agent systems with general linear and lipschitz nonlinear dynamics using distributed adaptive protocols. IEEE Trans. Autom. Control **58**(7), 1786–1791 (2012)
9. S. Hu, D. Yue, X. Xie, X. Chen, X. Yin, Resilient event-triggered controller synthesis of networked control systems under periodic dos jamming attacks. IEEE Trans. Cybern. **49**(12), 4271–4281 (2018)
10. C. Peng, D. Yue, M.-R. Fei, A higher energy-efficient sampling scheme for networked control systems over ieee 802.15. 4 wireless networks. IEEE Trans. Ind. Inf. **12**(5), 1766–1774 (2015)
11. L. Ding, Q.-L. Han, E. Sindi et al., Distributed cooperative optimal control of dc microgrids with communication delays. IEEE Trans. Ind. Inf. **14**(9), 3924–3935 (2018)
12. H. Rezaee, F. Abdollahi, Consensus problem in high-order multiagent systems with lipschitz nonlinearities and jointly connected topologies. IEEE Trans. Syst. Man Cybern. Syst. **47**(5), 741–748 (2017)
13. Z. Zuo, J. Zhang, Y. Wang, Adaptive fault-tolerant tracking control for linear and lipschitz nonlinear multi-agent systems. IEEE Trans. Ind. Electron. **62**(6), 3923–3931 (2014)
14. R.J. Adams, J.M. Buffington, A.G. Sparks, S.S. Banda, *Robust Multivariable Flight Control* (Springer Science and Business Media, 2012)

Part III
Distributed Cooperative Attitude Control of Multiple Rigid Body Systems

Chapter 6
Distributed Event-Triggered Attitude Cooperative Control of Multiple Rigid Body Systems

In this chapter, the distributed event-triggered cooperative attitude control of multiple groups of rigid bodies evolved on nonlinear manifold $SO(3)$ is investigated. The distributed event-triggered mechanism is introduced into the controller design to reduce the communication pressure. It is verified that, the orientations of rigid bodies in each group converge to a common value with zero angular velocities, and these common values in all groups achieve an anti-consensus configuration under the proposed intra- and inter-group cooperative control law. Simulation results show the effectiveness of the proposed method.

The rest of this chapter is organized as follows. Section 6.1 gives preliminaries and the problem formulation. Our main results are given in Sect. 6.2. In Sect. 6.3, numerical simulations are provided. Finally the chapter is concluded in Sect. 6.4.

6.1 Problem Formulation

In this chapter, the orientation of a rigid body in a three-dimensional Euclid space is represented by the element in the special orthogonal group $SO(3) = \{A \in R^{3\times3} | A^T A = AA^T = I, det(A) = 1\}$. The Lie algebra of $SO(3)$ is denoted as $so(3)$, which is the space of antisymmetric 3×3 matrices. The norm of $B \in R^{3\times3}$ is denoted by the Frobenius norm as $||B||^2 = tr(B^T B)$, and the scalar product of B and C in $R^{3\times3}$ is defined by $B \cdot C = tr(B^T C)$. Define the operator $s(\cdot) : R^{3\times3} \rightarrow R^9$ as $s(B) = (b_{11}, b_{12}, b_{13}, b_{21}, b_{22}, b_{23}, b_{31}, b_{32}, b_{33})^T$ for $B = (b_{ij}) \in R^{3\times3}$. Obviously, $tr(B^T C) = s^T(B)s(C)$ for any $B, C \in R^{3\times3}$.

Given a vector $a = (a_1, a_2, a_3)^T \in R^3$, the invertible operator $(\cdot)^\wedge$ is defined as follows

D. Yue et al., *Distributed Cooperative Control and Communication for Multi-agent Systems*, https://doi.org/10.1007/978-981-33-6718-0_6

$$(a)^\wedge = \begin{pmatrix} 0 & -a_3 & a_2 \\ a_3 & 0 & -a_1 \\ -a_2 & a_1 & 0 \end{pmatrix} \in so(3)$$

The inverse of operator $(\cdot)^\wedge$ is denoted by $(\cdot)^\vee : so(3) \to R^3$. By the definition of $(\cdot)^\wedge$, one can see that $a \wedge b = (a)^\wedge b$ for $a, b \in R^3$, where the symbol $\wedge$ in expression $a \wedge b$ denotes the vector product.

The following definition and lemmas will be used.

Definition 6.1 ([1]) The non-autonomous system of differential equations in R^n

$$\frac{d}{dt}x(t) = f(t, x) \tag{6.1}$$

is said to be asymptotically autonomous with limit equation

$$\frac{d}{dt}y(t) = g(y) \tag{6.2}$$

if

$$f(t, x) \to g(x), t \to \infty$$

uniformly holds on each compact subset of R^n. In this case, (6.2) is called as the autonomous limit system of (6.1).

Lemma 6.1 ([1]) *All solutions of an asymptotically autonomous system converge to the chain recurrent set of the limit system.*

Lemma 6.2 $tr(-(a)^\wedge(b)^\wedge) = 2a^T b$ *for any* $a, b \in R^3$

Proof Let $a = (a_1, a_2, a_3)^T$ and $b = (b_1, b_2, b_3)^T$, then $A = \begin{pmatrix} 0 & -a_3 & a_2 \\ a_3 & 0 & -a_1 \\ -a_2 & a_1 & 0 \end{pmatrix}$ and $B = \begin{pmatrix} 0 & -b_3 & b_2 \\ b_3 & 0 & -b_1 \\ -b_2 & b_1 & 0 \end{pmatrix}$. Therefore

$$tr(-(a)^\wedge(b)^\wedge) = 2(a_1b_1 + a_2b_2 + a_3b_3) = 2a^T b$$

Lemma 6.3 ([2]) *The gradient* $grad_{x_0, SO(3)} f(x)$ *along manifold* $SO(3)$ *of a function* $f(x) : SO(3) \to R$*, evaluated at* x_0*, can be defined as the vector resulting from the orthogonal projection of* $grad_{x_0, R^{3\times3}} f(x)$ *onto the tangent space* $T_{x_0}SO(3)$*, i.e.* $grad_{x_0, SO(3)} f(x) = \mathbf{Proj}_{T_{x_0}SO(3)}(grad_{x_0, R^{3\times3}} f(x))$.

Lemma 6.4 ([2]) *The orthogonal projection of* $B \in R^{3\times3}$ *onto the tangent space* $T_Q SO(3)$ *to* $SO(3)$ *at* Q *is* $Q(\frac{Q^T B}{2} - \frac{B^T Q}{2})$*, i.e.,* $\mathbf{Proj}_{T_Q SO(3)}(B) = Q(\frac{Q^T B}{2} - \frac{B^T Q}{2})$.

Given a network consisting of $\overline{n}$ rigid bodies, each rigid body has the attitude kinematics and dynamics as follows

$$\frac{d}{dt}Q_k = Q_k(\omega_k)^{\wedge} \tag{6.3}$$

$$J_k\frac{d}{dt}\omega_k = (J_k\omega_k) \wedge \omega_k + u_k, \qquad k = 1, 2, \ldots, \overline{n} \tag{6.4}$$

where $Q_k \in SO(3)$ represents the orientation of the kth rigid body, $\omega_k \in R^3$ denotes the angular velocity expressed in the body frame of the kth rigid body. $J_k \in R^{3\times3}$ is a constant positive-definite matrix denoting the moment of inertia matrix, and $u_k \in R^3$ is the control torque. (6.3) and (6.4) are also called Euler equations for rotational movements.

Remark 6.1 In order to study the attitude control in a convenient way, typical various attitude parameterizations, such as Euler angles, quaternions, Rodrigues parameters, are used [3]. However, all these parameterizations fail to represent the set of attitudes both globally and uniquely. In this chapter, the nonlinear manifold $SO(3)$ is used to represent the set of three-dimensional orientations in system (6.3)–(6.4). By the knowledge in [3], the manifold $SO(3)$ is the only method to represent the set of attitudes both globally and uniquely. Therefore, the system (6.3)–(6.4) has some advantages in representing the attitude of a rigid body over those attitude parameterizations methods.

The multi-group architecture is considered in this chapter. In each group, the rigid bodies interact with each other through the follower communication network, while the interaction between groups is implemented through the leader communication network. The control purpose is to synchronize the orientations of all rigid bodies in each group, and drive the synchronized orientations of all groups to an anti-consensus configuration. Moreover, the event-triggered strategy will be introduced to reduce the communication pressure. The tow types of configurations are defined as follows, and more details can be found in [4].

Definition 6.2 Consider a set of rigid bodies with orientations $Q_k \in SO(3), k = 1, \ldots, \overline{n}$. Synchronization is the configuration where $Q_j = Q_k, \forall j, k$.

Definition 6.3 Consider a set of rigid bodies with orientations $Q_k \in SO(3), k = 1, \ldots, \overline{n}$. Suppose that the rigid bodies are interconnected according to a fixed digraph of weighted adjacency matrix $A = (a_{jk})$. An anti-consensus configuration satisfies:

$$Q_k \in \underset{c\in SO(3)}{argmin}\big(trace(c^T \sum_{j=1}^{\overline{n}} a_{jk}Q_j)\big), \qquad k = 1, 2, \ldots, \overline{n}$$

Denote N as the set of all rigid bodies, which is divided into m groups G^1, $G^2, \ldots,$ G^m satisfying $G^1 \cup G^2 \cup \cdots \cup G^m = N$ and $G^i \cap G^j = \emptyset$ for $i \neq j$. For all rigid bodies in each group $G^i\,(i = 1, \ldots, m)$, there is one leader denoted as v_{l_i}, and the rest are flowers $v_1^i, v_2^i, \ldots, v_{|G^i|-1}^i$ where $|G^i|$ means the cardinal number of the set G^i. Obviously, $G^i = \{v_{l_i}, v_1^i, v_2^i, \ldots, v_{|G^i|-1}^i\}$. Two different types of communication graphs are concerned with the definitions given as follows:

(1) The **leader communication graph** $\mathcal{G}^L \triangleq \{L, E^L, A^L\}$ consists of a node set L defined as $L = \bigcup_{i=1,\ldots,m}\{v_{l_i}\}$, and a set of edges $E^L = \{(v_{l_i}, v_{l_j}) \in L \times L : v_{l_i} \in N_j^L\}$, where N_j^L denotes the neighboring set of rigid body v_{l_j} in graph $\mathcal{G}^L$ and v_{l_j} could get the information of those rigid bodies in N_j^L. $A^L = (a_{ij}^L) \in R^{|L|\times|L|}$ is the weighted adjacency matrix of graph $\mathcal{G}^L$ with nonnegative elements. $a_{ij}^L = 1$ if and only if $(v_{l_i}, v_{l_j}) \in E^L$, otherwise $a_{ij}^L = 0$, and it is assumed that $a_{ii}^L = 0$.

(2) The **follower communication graph** $\mathcal{G}^i \triangleq \{G^i\backslash\{v_{l_i}\}, E^i, A^i\}(i = 1, \ldots, m)$ consists of the node set $G^i\backslash\{v_{l_i}\}$ which denotes the set of all followers in group G^i, and a set of edges $E^i = \{(v_j^i, v_k^i) \in G^i\backslash\{v_{l_i}\} \times G^i\backslash\{v_{l_i}\} : v_j^i \in N_k^i\}$, where N_k^i denotes the neighboring set of rigid body v_k^i in graph $\mathcal{G}^i$ and v_k^i can get the information of those rigid bodies in N_k^i. $A^i = (a_{jk}^i) \in R^{(|G^i|-1|)\times(|G^i|-1|)}$ is the weighted adjacency matrix of graph $\mathcal{G}^i$. $a_{jk}^i = 1$ if and only if $(v_j^i, v_k^i) \in E^i$, otherwise $a_{jk}^i = 0$, and it is assumed that $a_{jj}^i = 0$. Define the matrix $D^i \triangleq diag\{d_1^i, \ldots, d_{(|G^i|-1|)}^i\}$ where $d_j^i \triangleq \sum_{v_k^i \in N_j^i} a_{jk}^i$, then the Laplacian matrix of graph $\mathcal{G}^i$ is given by $L^i = D^i - A^i$.

With the above multi-group architecture, the inter-group cooperative controller will be designed for all the leaders which implement the inter-group communication. This inter-group controller will drive the orientations of all leaders to an anti-consensus configuration. On the other hand, the intra-group cooperative controller constructed for the followers aims to synchronize the orientation of follower to that of the leader located in the same group. Combining the inter- and intra-group cooperative control laws can drive the orientations of the rigid bodies in each group to one common value, and make all these common values to an anti-consensus configuration. It should be mentioned that, both of the two types of controllers will be designed in a distributed way.

Moreover, as mentioned in the introduction, there always exist a mass of rigid bodies in each group and the intra-group communication topology is very complex, which implies that the conventional control law based on continuous information transmission will generate a serious intra-group communication pressure. To solve this problem, the event-triggered strategy will be introduced in the intra-group cooperative algorithms. Under this strategy, each follower only need to transmit its own information to its neighbors at some discrete triggered times, thus the event-triggered strategy can be expected an effective way in reducing the intra-group communication pressure.

6.2 Inter-Group Anti-consensus Algorithms

The leaders' distributed controllers will be designed to drive the orientations of leaders to an anti-consensus configuration with zero angular velocities.

The orientation, angular velocity, moment of inertia matrix and control torque of the leader v_{l_k} are represented as Q_k, ω_k, J_k and u_k respectively in this section. For the **leader communication graph** $\mathcal{G}^L$, define the function

$$V^L = \sum_{v_{l_k} \in L} \sum_{v_{l_j} \in L} a_{jk}^L tr(Q_j^T Q_k) \tag{6.5}$$

The following Lemma represents the relationship between the function V^L and the anti-consensus configuration. The proof can be found in [4].

Lemma 6.5 ([4]) *Assume that the* ***leader communication graph*** $\mathcal{G}^L$ *is undirected and connected, then a local minimum of* V^L *necessarily corresponds to an anti-consensus configuration.*

Based on Lemma 6.5 and Lyapunov function method, the following result is concluded.

Theorem 6.1 *Assuming that the leader communication graph* $\mathcal{G}^L$ *is undirected and connected, design the following control law for all the leaders as*

$$\begin{aligned} u_k = &-(J_k\omega_k) \wedge \omega_k + \alpha J_k \sum_{v_{l_j} \in N_k^L} \left[-(\omega_k)^\wedge Q_k^T Q_j + Q_k^T Q_j (\omega_j)^\wedge + (\omega_j)^\wedge Q_j^T Q_k \right. \\ &\left. - Q_j^T Q_k (\omega_k)^\wedge\right] + \beta\left[\alpha \sum_{v_{l_j} \in N_k^L} (Q_k^T Q_j - Q_j^T Q_k)^\vee - \omega_k\right] \quad (\forall v_{l_k} \in L) \end{aligned} \tag{6.6}$$

with the constants $\alpha < 0$ *and* $\beta > 0$*, then the orientations of all the leaders are driven to an anti-consensus configuration with zero angular velocities almost globally.*

Proof Define the variable

$$\omega_k^d = \alpha \sum_{v_{l_j} \in N_k^L} (Q_k^T Q_j - Q_j^T Q_k)^\vee \tag{6.7}$$

According to (6.3), we can get

$$\frac{d}{dt}\omega_k^d = \alpha \sum_{v_{l_j} \in N_k^L} \left[-(\omega_k)^\wedge Q_k^T Q_j + Q_k^T Q_j (\omega_j)^\wedge + (\omega_j)^\wedge Q_j^T Q_k - Q_j^T Q_k (\omega_k)^\wedge\right]$$

Therefore, the controller (6.6) can be rewritten as

$$u_k = -(J_k\omega_k) \wedge \omega_k + J_k \frac{d}{dt}\omega_k^d + \beta(\omega_k^d - \omega_k) \tag{6.8}$$

Define the function

$$W = V^L + \frac{\theta}{2} \sum_{v_{l_k} \in L} (\omega_k - \omega_k^d)^T J_k (\omega_k - \omega_k^d)$$

where V^L is defined in Eq. (6.5) and θ is a positive constant to be determined. The derivative of V^L with respect to t is

$$\frac{d}{dt}V^L = \sum_{v_{l_k} \in L} tr(grad_{Q_k,SO(3)}(V^L)^T \frac{dQ_k}{dt}) \tag{6.9}$$

where $grad_{Q_k,SO(3)}(V^L)$ is the gradient of V^L with respect to Q_k on manifold $SO(3)$. Since $grad_{Q_k,R^{3\times3}}(V^L) = 2\sum_{v_{l_j} \in L} a_{jk}^L Q_j$ for the undirected graph $\mathcal{G}^L$ based on [4], according to Lemma 6.2, we have

$$\begin{aligned} grad_{Q_k,SO(3)}(V^L) &= \mathbf{Proj}_{T_{Q_k}SO(3)}(grad_{Q_k,R^{3\times3}}(V_L))r \\ &= 2\mathbf{Proj}_{T_{Q_k}SO(3)}(\sum_{v_{l_j} \in L} a_{jk}^L Q_j) \\ &= 2\mathbf{Proj}_{T_{Q_k}SO(3)}(\sum_{v_{l_j} \in L} a_{jk}^L (Q_j - Q_k)) \end{aligned}$$

where the last equation holds since $\mathbf{Proj}_{T_{Q_k}SO(3)}(Q_k) = 0$.
According to Lemma 6.4, we have

$$\begin{aligned} \mathbf{Proj}_{T_{Q_k}SO(3)}(Q_j - Q_k) &= \frac{Q_k}{2}(Q_k^T(Q_j - Q_k) - (Q_j - Q_k)^T Q_k) \\ &= \frac{Q_k}{2}(Q_k^T Q_j - Q_j^T Q_k) \end{aligned}$$

thus

$$\begin{aligned} grad_{Q_k,SO(3)}(V^L) &= \sum_{v_{l_j} \in L} a_{jk}^L Q_k (Q_k^T Q_j - Q_j^T Q_k) \\ &= \frac{1}{\alpha} Q_k (\omega_k^d)^\wedge \end{aligned} \tag{6.10}$$

Substituting (6.3) and (6.10) into (6.9) yields

$$\begin{aligned}\frac{d}{dt}V^L &= \frac{1}{\alpha}\sum_{v_{l_k}\in L} tr((Q_k(\omega_k^d)^\wedge)^T Q_k(\omega_k)^\wedge)\\ &= \frac{1}{\alpha}\sum_{v_{l_k}\in L} tr(-(\omega_k^d)^\wedge(\omega_k)^\wedge)\\ &= \frac{2}{\alpha}\sum_{v_{l_k}\in L} (\omega_k^d)^T\omega_k \end{aligned} \tag{6.11}$$

where the last equation can be obtained by Lemma 6.2.

Using (6.4), (6.8), and (6.11), the derivative of W with respect to t is

$$\begin{aligned}\frac{d}{dt}W &= \sum_{v_{l_k}\in L}\left(\frac{2}{\alpha}(\omega_k^d)^T\omega_k - \theta\beta \parallel \omega_k^d - \omega_k \parallel^2\right)\\ &= \sum_{v_{l_k}\in L}\left(-\frac{1}{\alpha}(\parallel \omega_k^d - \omega_k \parallel^2 - \parallel \omega_k^d \parallel^2 - \parallel \omega_k \parallel^2) - \theta\beta \parallel \omega_k^d - \omega_k \parallel^2\right)\\ &= \sum_{v_{l_k}\in L}\left(-\left(\frac{1}{\alpha}+\theta\beta\right) \parallel \omega_k^d - \omega_k \parallel^2 + \frac{1}{\alpha}(\parallel \omega_k^d \parallel^2 + \parallel \omega_k \parallel^2)\right)\end{aligned}$$

Obviously, choosing $\theta > -\frac{1}{\alpha\beta} > 0$ makes $\frac{d}{dt}W \leqslant 0$, thus the angular velocity ω_k and the variable ω_k^d for all $v_{l_k} \in L$ converge to zero by Lasalle invariance principle. Furthermore, by the help of (6.11), $\omega_k^d = 0$ implies $\frac{d}{dt}V^L = 0$. Since the stable critical points must be minima of W, hence minima of V^L, and this corresponds to an anti-consensus configuration by Lemma 6.5. This concludes the proof.

6.3 Intra-Group Event-Triggered Synchronization Algorithms

The leaders' cooperative control law (6.6) can be implemented by using the sampled-data control method in practice. As long as the sampling period is selected as small enough, the discretized implementation of (6.6) can guarantee the similar dynamical performance of the closed-loop with (6.6). Since the number of groups (i.e. the number of leaders) is often small, the inter-group communication pressure is not heavy. We do not pay much attention to the method on reducing the communication pressure between the leaders. However, there usually exist a mass of rigid bodies and a complex intra-group communication topology in each group in applications, which leads to a serious problem in communication pressure.

The event-triggered strategy will be introduced in the design of the intra-group cooperative control law. Under this strategy, each rigid body only need to transmit its own information to its neighbors at some discrete triggered times in order to reduce the intra-group communication pressure. In this section, only the intra-group cooperative control law in group G^i is considered, and the control laws in other groups can be obtained by a similar analysis. The orientation, angular velocity and control torque of the follower v_k^i are labeled as Q_k, ω_k and u_k respectively. Denoting the monotonically increasing sequence $t_0^k, t_1^k, \ldots, t_n^k, \ldots$ as the triggered times of the follower $v_k^i (k = 1, 2, \ldots, |G^i| - 1)$, we propose the following intra-group event-triggered cooperative controller for the follower $v_k^i (\forall v_k^i \in G^i)$

$$u_k(t) = -\sigma(Q_k(t)^T Y_k(t) - Y_k(t)^T Q_k(t))^\vee - \gamma\omega_k(t) \tag{6.12}$$

$$\frac{d}{dt}Y_k(t) = \sum_{v_j^i \in N_k^i} (Y_j(t_{n'(t)}^j) - Y_k(t_n^k)) + b_k^i(Q_{l_i}(t_{n'(t)}^{l_i}) - Y_k(t)), t \in [t_n^k, t_{n+1}^k) \tag{6.13}$$

where $\sigma < 0$, $\gamma > 0$ and $b_k^i \in \{0, 1\}$ are some constants, $Y_k(t) \in R^{3\times3}$ is the auxiliary variable of the follower v_k^i, $t_{n'(t)}^j$ is the latest event-triggered time of the follower v_j^i at time t, and $Q_{l_i}(t_{n'(t)}^{l_i})$ is the latest triggered data about the orientation of the leader v_{l_i} in the group G^i at time t.

Define the measurement error as

$$e_k(t) = Y_k(t) - Y_k(t_n^k), \qquad t \in [t_n^k, t_{n+1}^k)$$

and suppose the last triggered time of the follower v_k^i is t_n^k. At time $t > t_n^k$, if

$$\| e_k(t) \| > \frac{\sum_{v_j^i \in N_k^i} \| Y_k(t_n^k) - Y_j(t_{n'(t)}^j) \|^2}{2(\| \sum_{v_j^i \in N_k^i} (Y_k(t_n^k) - Y_j(t_{n'(t)}^j)) \| + \xi)} \tag{6.14}$$

where $\xi > 0$ is a constant, then the auxiliary variable of v_k^i will be sampled and transmitted to the rigid bodies in N_k^i, and denote the time t as t_{n+1}^k.

Moreover, suppose the last triggered time of the leader v_{l_i} is $t_n^{l_i}$. At time $t > t_n^{l_i}$, if

$$t - t_n^{l_i} = T^{l_i} \tag{6.15}$$

where $T^{l_i} > 0$ is a constant, then the leader's orientation Q_{l_i} will be sampled and transmitted to the followers which can receive the information of v_{l_i}, and denote the time t as $t_{n+1}^{l_i}$.

Remark 6.2 It should be mentioned that, each follower rigid body v_k^i only need to transmit its own information to its neighbors in N_k^i at some discrete time when the triggering condition (6.14) is satisfied. Meanwhile, the leader rigid body v_{l_i} periodically transmits its own information to the followers with periodic time T^{l_i}. Compared with the conventional control law with continuous information transmission, it can be expected a lower intra-group communication pressure under the event-triggered cooperative control strategy (6.12)–(6.15).

Lemma 6.6 *Assuming that the follower communication graph $\mathcal{G}^i$ is undirected and connected, and at least one b_k^i equals to 1 for $k = 1, 2, \ldots, |G^i \backslash \{v_{l_i}\}|$, consider the system*

$$\frac{d}{dt}Y_k(t) = \sum_{v_j^i \in N_k^i} (Y_j(t_{n'(t)}^j) - Y_k(t_n^k)) + b_k^i(\overline{Y}^i - Y_k(t)), \; t \in [t_n^k, t_{n+1}^k) \tag{6.16}$$

where $k = 1, 2, \ldots, |G^i \backslash \{v_{l_i}\}|$, the variable $Y_k(t) \in R^{3\times 3}$ and the constant matrix $\overline{Y}^i \in R^{3\times 3}$, then $Y_k(t)$ converges to $\overline{Y}^i$ globally under the event-triggered mechanism (6.14). Moreover, the inter-event time interval of the triggering condition (6.14) is lower bounded by some positive value.

Proof Defining the variable $\delta_k(t) = Y_k(t) - \overline{Y}^i$, the derivative of $\delta_k(t)$ with respect to time t is

$$\begin{aligned}\frac{d}{dt}\delta_k(t) &= \frac{d}{dt}Y_k(t)\\ &= \sum_{v_j^i \in N_k^i} (Y_j(t_{n'(t)}^j) - Y_k(t_n^k)) + b_k^i(\overline{Y}^i - Y_k(t))\\ &= \sum_{v_j^i \in N_k^i} (\delta_j(t_{n'(t)}^j) - \delta_k(t_n^k)) - b_k^i \delta_k(t),\end{aligned} \tag{6.17}$$

Constructing the following nonnegative function

$$V = \frac{1}{2} \sum_{v_k^i \in G^i \backslash \{v_{l_i}\}} \| \delta_k(t) \|^2$$

according to (6.17), the derivative of V with respect to time t is

$$\begin{aligned}\frac{d}{dt}V &= \sum_{v_k^i \in G^i \backslash \{v_{l_i}\}} \delta_k(t) \cdot \frac{d}{dt}\delta_k(t)\\ &= \sum_{v_k^i \in G^i \backslash \{v_{l_i}\}} \delta_k(t) \cdot \left(\sum_{j \in N_k^i} (\delta_j(t_{n'(t)}^j) - \delta_k(t_n^k)) - b_k^i \delta_k(t) \right)\\ &= \sum_{v_k^i \in G^i \backslash \{v_{l_i}\}} \delta_k(t) \cdot \sum_{v_j^i \in N_k^i} (\delta_j(t_{n'(t)}^j) - \delta_k(t_n^k))\\ &- \sum_{v_k^i \in G^i \backslash \{v_{l_i}\}} b_k^i \| \delta_k(t) \|^2\end{aligned} \tag{6.18}$$

Since $\delta_k(t) = Y_k(t) - \overline{Y}^i = e_k(t) + Y_k(t_n^k) - \overline{Y}^i = e_k(t) + \delta_k(t_n^k)$, we have

$$\begin{aligned}
&\sum_{v_k^i \in G^i \setminus \{v_{l_i}\}} \delta_k(t) \cdot \sum_{v_j^i \in N_k^i} (\delta_j(t_{n'(t)}^j) - \delta_k(t_n^k)) \\
&= \sum_{v_k^i \in G^i \setminus \{v_{l_i}\}} (e_k(t) + \delta(t_n^k)) \cdot \sum_{v_j^i \in N_k^i} (\delta_j(t_{n'(t)}^j) - \delta_k(t_n^k)) \\
&= \sum_{v_k^i \in G^i \setminus \{v_{l_i}\}} \delta(t_n^k) \cdot \sum_{v_j^i \in N_k^i} (\delta_j(t_{n'(t)}^j) - \delta_k(t_n^k)) \\
&+ \sum_{v_k^i \in G^i \setminus \{v_{l_i}\}} e_k(t) \cdot \sum_{v_j^i \in N_k^i} (\delta_j(t_{n'(t)}^j) - \delta_k(t_n^k))
\end{aligned} \tag{6.19}$$

Moreover, by the fact that the graph $\mathcal{G}^i$ is undirected, we get

$$\begin{aligned}
&\sum_{v_k^i \in G^i \setminus \{v_{l_i}\}} \delta(t_n^k) \cdot \sum_{v_j^i \in N_k^i} (\delta_j(t_{n'(t)}^j) - \delta_k(t_n^k)) \\
&= \sum_{v_k^i \in G^i \setminus \{v_{l_i}\}} \sum_{v_j^i \in N_k^i} (\delta_k(t_n^k) \cdot \delta_j(t_{n'(t)}^j) - \delta_k(t_n^k) \cdot \delta_k(t_n^k)) \\
&= \sum_{v_k^i \in G^i \setminus \{v_{l_i}\}} \sum_{v_j^i \in N_k^i} (\delta_k(t_n^k) \cdot \delta_j(t_{n'(t)}^j) - \frac{1}{2}\delta_k(t_n^k) \cdot \delta_k(t_n^k) - \frac{1}{2}\delta_j(t_{n'(t)}^j) \cdot \delta_j(t_{n'(t)}^j)) \\
&= -\frac{1}{2} \sum_{v_k^i \in G^i \setminus \{v_{l_i}\}} \sum_{v_j^i \in N_k^i} \| \delta_j(t_{n'(t)}^j) - \delta_k(t_n^k) \|^2
\end{aligned} \tag{6.20}$$

Substituting (6.19) and (6.20) into (6.18) yields

$$\begin{aligned}
\frac{d}{dt}V \leqslant & \sum_{v_k^i \in G^i \setminus \{v_{l_i}\}} \Bigg(-\frac{1}{2} \sum_{v_j^i \in N_k^i} \| \delta_j(t_{n'(t)}^j) - \delta_k(t_n^k) \|^2 + \| e_k(t) \| \\
& \cdot \| \sum_{v_j^i \in N_k^i} (\delta_j(t_{n'(t)}^j) - \delta_k(t_n^k)) \| \Bigg) - \sum_{v_k^i \in G^i \setminus \{v_{l_i}\}} b_k^i(t) \| \delta_k(t) \|^2
\end{aligned} \tag{6.21}$$

Based on the triggering condition (6.14), we can get

$$\begin{aligned}
\| e_k(t) \| &\leqslant \frac{\sum_{v_j^i \in N_k^i} \| Y_k(t_n^k) - Y_j(t_{n'(t)}^j) \|^2}{2(\| \sum_{v_j^i \in N_k^i} (Y_k(t_n^k) - Y_j(t_{n'(t)}^j)) \| + \xi)} \\
&= \frac{\sum_{v_j^i \in N_k^i} \| \delta_k(t_n^k) - \delta_j(t_{n'(t)}^j) \|^2}{2(\| \sum_{v_j^i \in N_k^i} (\delta_k(t_n^k) - \delta_j(t_{n'(t)}^j)) \| + \xi)}
\end{aligned}$$

which further implies

$$
\begin{aligned}
&-\frac{1}{2}\sum_{v_j^i \in N_k^i} \| \delta_j(t_{n'(t)}^j) - \delta_k(t_n^k) \|^2 + \| e_k(t) \| \cdot \| \sum_{v_j^i \in N_k^i} \left(\delta_j(t_{n'(t)}^j) - \delta_k(t_n^k)\right) \\
&\leqslant -\xi e_k(t) \|
\end{aligned} \tag{6.22}
$$

Substituting (6.22) into (6.21) yields

$$
\frac{d}{dt}V \leqslant - \sum_{v_k^i \in G^i \backslash \{v_{l_i}\}} \xi \| e_k(t) \| - \sum_{v_k^i \in G^i \backslash \{v_{l_i}\}} b_k^i \| \delta_k(t) \|^2 \leqslant 0 \tag{6.23}
$$

It should be mentioned that, the Eq. (6.18) can be rewritten as

$$
\begin{aligned}
\frac{d}{dt}V =& \sum_{v_k^i \in G^i \backslash \{v_{l_i}\}} \delta_k(t) \cdot \sum_{v_j^i \in N_k^i} \left(\delta_j(t) - \delta_k(t) - (e_j(t) - e_k(t))\right) - \sum_{v_k^i \in G^i \backslash \{v_{l_i}\}} b_k^i \| \delta_k(t) \|^2 \\
=& \sum_{v_k^i \in G^i \backslash \{v_{l_i}\}} s^T(\delta_k(t)) \sum_{v_j^i \in N_k^i} s(\delta_j(t) - \delta_k(t)) - \sum_{v_k^i \in G^i \backslash \{v_{l_i}\}} s^T(\delta_k(t)) \sum_{v_j^i \in N_k^i} s(e_j(t) - e_k(t)) \\
&- \sum_{v_k^i \in G^i \backslash \{v_{l_i}\}} b_k^i \| \delta_k(t) \|^2 \\
=& -s^T(\delta(t))(L^i \otimes I_9)s(\delta(t)) + s^T(\delta(t))(L^i \otimes I_9)s(e(t)) - \sum_{v_k^i \in G^i \backslash \{v_{l_i}\}} b_k^i \| \delta_k(t) \|^2
\end{aligned} \tag{6.24}
$$

where the operator $s(\cdot)$ is defined in Sect. 6.1, $s(\delta(t)) \triangleq (s(\delta_1(t))^T, s(\delta_2(t))^T, \ldots, s(\delta_{|G^i \backslash \{v_{l_i}\}|}(t))^T)^T$, L^i is the Laplacian matrix of graph $\mathcal{G}^i$ which is semi-positive definite, $s(e(t)) \triangleq (s(e_1(t))^T, s(e_2(t))^T, \ldots, s(e_{|G^i \backslash \{v_{l_i}\}|}(t))^T)^T$, I_9 is the nine-order identity matrix and the notation $\otimes$ means the Kronecker product. Since $\frac{d}{dt}V \leqslant 0$, according to (6.23), (6.24) and Lasalle invariance principle, the solution of the system (6.16) converges into the following set

$$
\begin{aligned}
S \triangleq \{&(Y_1, Y_2, \ldots, Y_{|G^i \backslash \{v_{l_i}\}|}) : (L^i \otimes I_9)s(\delta) = 0, \sum_{v_k^i \in G^i \backslash \{v_{l_i}\}} b_k^i \| \delta_k \|^2 = 0, \\
&\sum_{v_k^i \in G^i \backslash \{v_{l_i}\}} \xi \| e_k \| = 0\}
\end{aligned}
$$

This implies that $L^i s_1(\delta) = L^i s_2(\delta) = \cdots = L^i s_9(\delta) = 0$ in set S, where $s_1(\delta)$, $s_2(\delta), \ldots, s_9(\delta) \in R^{|G^i \backslash \{v_{l_i}\}|}$ are the stack vectors of the nine coefficients of all $s(\delta_k) \in R^9$ respectively. Since the graph $\mathcal{G}^i$ is undirected and connected, it could be obtained that $s_1(\delta) = a_1 \mathbf{1}^{|G^i \backslash \{v_{l_i}\}|}$, $s_2(\delta) = a_2 \mathbf{1}^{|G^i \backslash \{v_{l_i}\}|}$, ...,$s_9(\delta) = a_9 \mathbf{1}^{|G^i \backslash \{v_{l_i}\}|}$ where $a_1, a_2, \ldots a_9$ are nine constants [5]. Therefore, for any rigid body $v_k^i \in G^i \backslash \{v_{l_i}\}$, we have $s(\delta_k) = (a_1, a_2, \ldots a_9)^T \triangleq s^*$ in set S. Moreover, since there is at least one $b_k^i = 1$, $\sum_{v_k^i \in G^i \backslash \{v_{l_i}\}} b_k^i \parallel \delta_k \parallel^2 = 0$ implies at least one $\delta_k = 0$, and thus $s^* = 0$. Therefore, $\delta_k = 0$ for all $v_k^i \in G^i \backslash \{v_{l_i}\}$, and this implies $Y_k = \overline{Y}^i$ for all $v_k^i \in G^i \backslash \{v_{l_i}\}$ in set S.

Consider the inter-event time interval $t_{n+1}^k - t_n^k$ of the triggering condition (6.14). For $t \in [t_n^k, t_{n+1}^k)$, we have $\frac{d}{dt} \parallel e_k(t) \parallel \leqslant \parallel \frac{d}{dt} e_k(t) \parallel = \parallel \frac{d}{dt} Y_k(t) \parallel$. Since the continuous differential function $Y_k(t)$ converges to $\overline{Y}^i$, we get $\frac{d}{dt} \parallel e_k(t) \parallel \leqslant M$ for some constant $M > 0$. This implies $\parallel e_k(t) \parallel \leqslant M(t - t_n^k)$ since $e_k(t_n^k) = 0$. Furthermore, the triggering condition (6.14) signifies that $lim_{t \to t_{n+1}^k} \parallel e_k(t) \parallel = \frac{\sum_{v_j^i \in N_k^i} \|Y_k(t_n^k) - Y_j(t^j_{n'(t_{n+1}^k)})\|^2}{2(\|\sum_{v_j^i \in N_k^i} (Y_k(t_n^k) - Y_j(t^j_{n'(t_{n+1}^k)}))\| + \xi)}$. The above results show that $t_{n+1}^k - t_n^k \geqslant \frac{\sum_{v_j^i \in N_k^i} \|Y_k(t_n^k) - Y_j(t^j_{n'(t_{n+1}^k)})\|^2}{2M(\|\sum_{v_j^i \in N_k^i} (Y_k(t_n^k) - Y_j(t^j_{n'(t_{n+1}^k)}))\| + \xi)}$, which ends the proof.

The following lemma will be used in Theorem 6.2, the proof can be found in [4] and is thus omitted here.

Lemma 6.7 ([4]) *The local maxima of function $f(Q) = trace(Q^T Y)$ over $Q \in SO(3)$, with Y fixed in $R^{3\times3}$, are all global maxima.*

Combining the inter-group algorithms for the leaders and intra-group algorithms for the followers, the main result is presented as follows.

Theorem 6.2 *Suppose that the follower communication graph $\mathcal{G}^i (i = 1, \ldots, m)$ and the leader communication graph $\mathcal{G}^L$ are undirected and connected, the leader rigid body $v_{l_i} (i = 1, \ldots, m)$ is in control of (6.6) with the triggering mechanism (6.15), and the follower rigid body $v_k^i (k = 1, 2, \ldots, |G^i| - 1, i = 1, 2, \ldots m)$ is in control of (6.12)–(6.13) with the triggering mechanism (6.14) where at least one $b_k^i = 1$ for each i, then the orientations of the leaders converge to an anti-consensus configuration, and the orientations of the followers in each group converge to that of the leader located in the same group almost globally. Moreover, the angular velocity of each rigid body converges to zero.*

Proof From Theorem 6.1, it is known that the orientations of the leaders in the set L converge to an anti-consensus configuration with zero angular velocities. Denote $\overline{Q}_{l_i}$ as the converged orientation of the leader v_{l_i}, then replacing $Q_{l_i}(t^{l_i}_{n'(t)})$ with $\overline{Q}_{l_i}$ in (6.13) obtains the autonomous limit system of (6.13) as follows

$$\frac{d}{dt} Y_k(t) = \sum_{v_j^i \in N_k^i} (Y_j(t^j_{n'(t)}) - Y_k(t_n^k)) + b_k^i(\overline{Q}_{l_i} - Y_k(t)), t \in [t_n^k, t_{n+1}^k) \quad (6.25)$$

In term of Lemma 6.6, the solution of system (6.25) converges to $\overline{Q}_{l_i}$. Therefore, the chain recurrent set of system (6.25) reduces to the node set $\{\overline{Q}_{l_i}\}$. This implies that the solution $Y_k(t)$ of system (6.13) converges to $\overline{Q}_{l_i}$ by Lemma 6.1. Hence the autonomous limit system of the closed-loop system with substituting (6.12)–(6.13) into (6.3)–(6.4) can be obtained as

$$\frac{d}{dt}Q_k = Q_k(\omega_k)^\wedge \tag{6.26}$$

$$J_k\frac{d}{dt}\omega_k = (J_k\omega_k)\wedge\omega_k + -\sigma(Q_k(t)^T\overline{Q}_{l_i} - \overline{Q}_{l_i}^T Q_k(t))^\vee - \gamma\omega_k(t) \tag{6.27}$$

By a similar proof in Theorem 2 of [6], it can be shown from (6.26)–(6.27) that, the variable ω_k converges to 0 and Q_k converges to the maximizer of the function $trace(Q_k^T\overline{Q}_{l_i})$ almost globally. Obviously, $Q_k = \overline{Q}_{l_i}$ is the global maxima of the function $trace(Q_k^T\overline{Q}_{l_i})$ over $Q_k \in SO(3)$. Therefore, it can be concluded that $Q_k = \overline{Q}_{l_i}$ is the unique maximizer by Lemma 6.7. Moreover, the function $trace(Q_k^T\overline{Q}_{l_i})$ has the unique maximizer $Q_k = \overline{Q}_{l_i}$ and three unstable points for each constant $\overline{Q}_{l_i}$ ([4]). Similar to the proof of Theorem 2 in [6], all solutions $Q_k(k = 1, 2, \ldots, |G^i\backslash\{v_{l_i}\}|)$ starting outside these three points and their stable manifolds converge to $\overline{Q}_{l_i}$, which ends the proof.

6.4 Simulation Example

Consider twelve rigid bodies as shown in Fig. 6.1, where l_1, l_2, l_3 are regarded as leaders, and the others are followers. All the rigid bodies are divided into three groups $G^1 = \{l_1, 1, 2, 3, 4\}$, $G^2 = \{l_2, 5, 6, 7\}$ and $G^3 = \{l_3, 8, 9\}$. The communication graph can be found in Fig. 6.1, where the dash lines indicate the inter-group communication and the solid lines denote the intra-group communication. Obviously, all the leader communication graph $\mathcal{G}^L$ and the follower communication graph $\mathcal{G}^i(i = 1, 2, 3)$ are undirected and connected, and at least one follower in G^i can

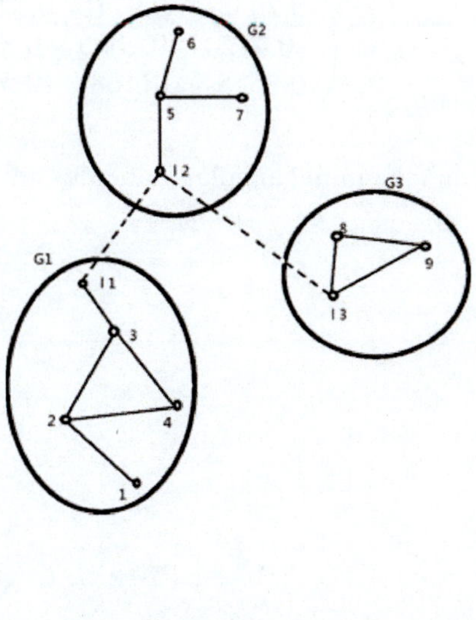

Fig. 6.1 Communication topology of rigid bodies

receive the information of the leader l_i for $i = 1, 2, 3$. Each rigid body's kinematics and dynamics are represented in (6.3)–(6.4). For simplicity and without loss of generality, the moment of inertia matrix of each rigid body is set as identity matrix. The initial orientations of all rigid bodies are

$$Q_{l_1}(0) = \begin{pmatrix} -0.2697 & 0.6604 & 0.7008 \\ -0.3539 & 0.6088 & -0.7100 \\ 0.8955 & 0.4395 & -0.0696 \end{pmatrix}; \; Q_{l_2}(0) = \begin{pmatrix} -0.7107 & 0.6199 & 0.3326 \\ 0.0160 & 0.4869 & -0.8733 \\ 0.7033 & 0.6153 & 0.3559 \end{pmatrix};$$

$$Q_{l_3}(0) = \begin{pmatrix} -0.2892 & -0.6915 & 0.6619 \\ 0.8873 & -0.4531 & -0.0857 \\ 0.3592 & 0.5626 & 0.7446 \end{pmatrix}; \; Q_1(0) = \begin{pmatrix} 0.4972 & -0.0586 & 0.8656 \\ -0.0862 & -0.9961 & -0.0179 \\ -0.8633 & 0.0657 & 0.5003 \end{pmatrix};$$

$$Q_2(0) = \begin{pmatrix} -0.5959 & 0.1790 & 0.7828 \\ 0.6916 & -0.3809 & 0.6136 \\ 0.4081 & 0.9071 & 0.1032 \end{pmatrix}; \; Q_3(0) = \begin{pmatrix} -0.7778 & -0.2408 & 0.5805 \\ 0.1850 & 0.7950 & 0.5777 \\ 0.6006 & -0.5568 & 0.5738 \end{pmatrix};$$

$$Q_4(0) = \begin{pmatrix} 0.5295 & 0.3610 & 0.7676 \\ 0.4690 & -0.8786 & 0.0897 \\ -0.7069 & -0.3125 & 0.6346 \end{pmatrix}; \; Q_5(0) = \begin{pmatrix} 0.6683 & -0.6241 & -0.4048 \\ 0.6722 & 0.2737 & 0.6879 \\ 0.3185 & 0.7318 & -0.6025 \end{pmatrix};$$

$$Q_6(0) = \begin{pmatrix} 0.2252 & -0.7104 & 0.6668 \\ -0.6373 & -0.6251 & -0.4507 \\ 0.7370 & -0.3235 & -0.5935 \end{pmatrix}; \; Q_7(0) = \begin{pmatrix} -0.3396 & 0.5484 & -0.7642 \\ 0.3289 & -0.6919 & -0.6427 \\ -0.8812 & -0.4696 & 0.0546 \end{pmatrix};$$

$$Q_8(0) = \begin{pmatrix} 0.2035 & -0.9715 & 0.1219 \\ -0.4724 & -0.2065 & -0.8568 \\ 0.8575 & 0.1168 & -0.5010 \end{pmatrix}; \; Q_9(0) = \begin{pmatrix} -0.0463 & 0.7723 & 0.6335 \\ -0.2538 & 0.6043 & -0.7552 \\ 0.9662 & 0.1957 & -0.1680 \end{pmatrix};$$

and the initial angular velocities are

$\omega_{1_1}(0) = (0.5406, 0.0193, 0.4139)^T$; $\omega_{l_2}(0) = (0.5823, -0.3387, 0.8274)^T$;
$\omega_{1_3}(0) = (0.1534, -0.4242, -0.4663)^T$; $\omega_1(0) = (-0.8056, 0.1190, -0.9138)^T$;
$\omega_2(0) = (0.5153, 0.3193, 0.1989)^T$; $\omega_3(0) = (-0.5460, 0.4274, 0.4490)^T$;
$\omega_4(0) = (-0.3116, -0.9345, 0.6872)^T$; $\omega_5(0) = (-0.1372, 0.8213, -0.6363)^T$;
$\omega_6(0) = (-0.4724, -0.7089, -0.7279)^T$; $\omega_7(0) = (0.7386, 0.1594, 0.0997)^T$;
$\omega_8(0) = (-0.7101, 0.7061, 0.2441)^T$; $\omega_9(0) = (-0.2981, 0.0265, -0.1964)^T$;

Moreover, the initial values of the followers' auxiliary variables are set as

$$Y_1(0) = \begin{pmatrix} -0.1254 & -0.3934 & -0.2614 \\ 0.2252 & -0.9141 & 0.3227 \\ 0.4750 & 0.6710 & 0.7914 \end{pmatrix}; Y_2(0) = \begin{pmatrix} 0.5765 & 0.7475 & -0.0113 \\ 0.7058 & -0.3323 & -0.9376 \\ -0.0288 & -0.5894 & 0.6551 \end{pmatrix};$$

$$Y_3(0) = \begin{pmatrix} -0.5818 & -0.3372 & -0.8251 \\ -0.4551 & 0.2061 & -0.3822 \\ 0.5516 & -0.6320 & -0.5382 \end{pmatrix}; Y_4(0) = \begin{pmatrix} -0.4896 & 0.1253 & -0.3245 \\ 0.8094 & 0.7943 & 0.2377 \\ 0.5346 & -0.2638 & -0.8262 \end{pmatrix};$$

$$Y_5(0) = \begin{pmatrix} 0.8588 & -0.1283 & 0.0170 \\ 0.5514 & -0.1064 & 0.0215 \\ -0.0264 & -0.3873 & 0.6353 \end{pmatrix}; Y_6(0) = \begin{pmatrix} 0.5897 & 0.6232 & 0.8780 \\ 0.2886 & 0.0657 & 0.7519 \\ -0.2428 & -0.2985 & 0.1003 \end{pmatrix};$$

$$Y_7(0) = \begin{pmatrix} 0.2450 & -0.3975 & 0.6886 \\ 0.1741 & -0.0582 & -0.6105 \\ -0.5845 & -0.5390 & -0.5482 \end{pmatrix}; Y_8(0) = \begin{pmatrix} -0.6586 & -0.3778 & -0.6304 \\ -0.5447 & 0.8468 & 0.8098 \\ -0.1286 & -0.1396 & 0.9595 \end{pmatrix};$$

$$Y_9(0) = \begin{pmatrix} -0.1223 & -0.1826 & 0.2057 \\ -0.7778 & 0.1898 & 0.4224 \\ -0.4839 & -0.4756 & -0.5565 \end{pmatrix};$$

Under the inter-group algorithm (6.6) with the triggered scheme (6.15) for the leaders l_1, l_2, l_3 and intra-group algorithm (6.12)–(6.13) with the triggered scheme (6.14) for the followers where the parameters are set as $T^{l_i} = 0.05 (i = 1, 2, 3)$, $\alpha = -2$, $\beta = 3$, $\sigma = -2$, $\gamma = 4$ and $\xi = 0.1$, the simulation results show that the orientations of the three leaders converge to an anti-consensus configuration, and those of the followers in each group converge to the orientation of the leader located in this group. Moreover, the angular velocities of all rigid bodies converge to zero.

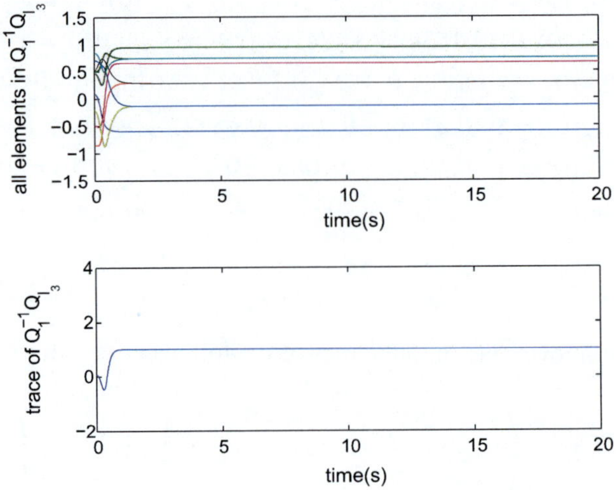

Fig. 6.2 Time responses of relative orientation between rigid bodies l_1 and l_3

Due to the space limitations, we only present the simulation results of some selected rigid bodies here.

Figure 6.2 shows the time responses of relative orientation between the leaders l_1 and l_3. By simply calculation, $X_1 = X_2$ if and only if $trace(X_1^{-1}X_2) = 3$ for $X_1, X_2 \in SO(3)$. It could be seen that, these two leaders' orientations do not converge to the same value since $trace(Q_{l_1}^{-1}Q_{l_3})$ does not converge to 3. This reflects the anti-consensus configuration to which the leaders' orientations converge. In order to show the advantage of the event-triggered controller (6.12)–(6.13), a type of non-triggered conventional continuous controller is designed as

$$u_k(t) = -\sigma(Q_k(t)^T Y_k(t) - Y_k(t)^T Q_k(t))^\vee - \gamma\omega_k(t) \tag{6.28}$$

$$\frac{d}{dt}Y_k(t) = \sum_{v_j^i \in N_k^i} (Y_j(t) - Y_k(t)) + b_k^i(Q_{l_i}(t) - Y_k(t)) \tag{6.29}$$

whose execution needs continuous information transmission. Figure 6.3a, b show the time responses of the relative orientation between the rigid body 1 and the leader l_1 under the event-triggered controller (6.12)–(6.13) and the continuous controller (6.28)–(6.29) respectively. Compared with the event-triggered controller there is no major system performance improvement as shown in Fig. 6.3, although the continuous information transmission method is used in the continuous controller. This illustrates the efficiency and superiority of event-triggered control method in term of load on the communication network. Figure 6.4a, b illustrate the the angular velocity variations of the follower rigid body 1 and the leaders l_1 and l_3 under the event-triggered controller and the continuous controller respectively. It can be seen that all the angular velocities converge to zero. Moreover, the two type of controllers share the similar control performance, which reflects the effectiveness and advantage of the event-triggered control strategy. Figure 6.5 shows the broadcast periods of aux-

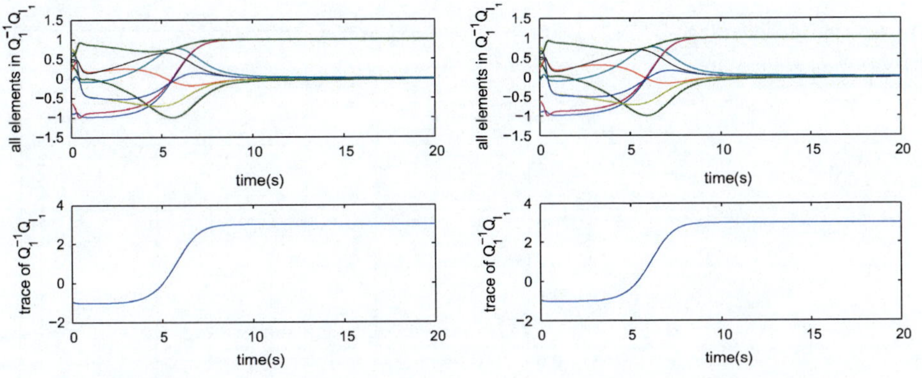

(a) Under event-triggered controller (b) Under continuous controller

Fig. 6.3 Time responses of relative orientation between rigid bodies 1 and l_1

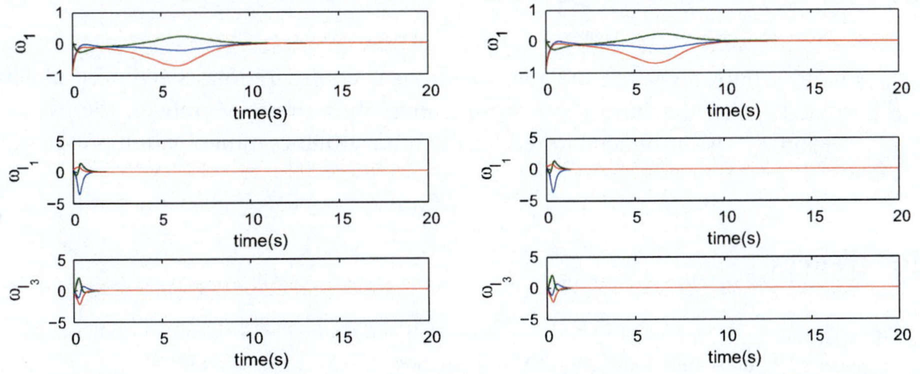

(a) Under event-triggered controller (b) Under continuous controller

Fig. 6.4 Time responses of angular velocities of rigid bodies 1, l_1 and l_3

iliary variable Y_1 of rigid body 1. The x-coordinates of circles in this plot signify the event-triggered time, and the y-coordinates denote the elapsed time since the last event triggering. Figure 6.5 shows that, the broadcast periods are not equivalent and the low bound of inter-event time interval is $0.228s$. It demonstrates the ability of event-triggering in adjusting broadcasting periods in response to variations in the system's states.

6.5 Conclusion

The problem of distributed cooperative attitude control of multiple groups of rigid bodies on manifold $SO(3)$ has been studied in this chapter. Under the multi-group architecture, the control objective of multiple attitude agreements, where the orientations of rigid bodies in each group converge to a common value and these common

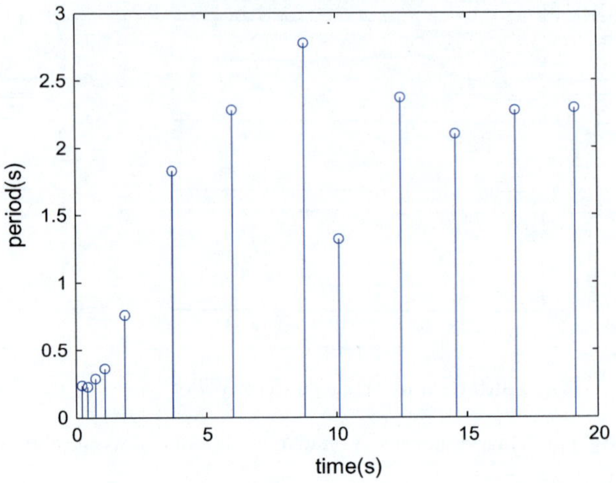

Fig. 6.5 the broadcasting periods of Y_1 about rigid body 1

values in all groups converge to an anti-consensus configuration, is achieved under the designed inter- and intra-group cooperative algorithms. Moreover, the event-triggered strategy is introduced to reduce the intra-group communication pressure.

References

1. K. Mischaikow, H. Smith, H.R. Thieme, Asymptotically autonomous semiflows: chain recurrence and lyapunov functions. Trans. Am. Math. Soc. **347**(5), 1669–1685 (1995)
2. A. Sarlette, Geometry and symmetries in coordination control, Ph.D. dissertation, Universit de Lige, 2009
3. N. Chaturvedi, A.K. Sanyal, N.H. McClamroch et al., Rigid-body attitude control. IEEE Control Syst. **31**(3), 30–51 (2011)
4. A. Sarlette, R. Sepulchre, Consensus optimization on manifolds. SIAM J. Control Optim. **48**(1), 56–76 (2009)
5. R. Olfati-Saber, R.M. Murray, Consensus problems in networks of agents with switching topology and time-delays. IEEE Trans. Autom. Control **49**(9), 1520–1533 (2004)
6. A. Sarlette, R. Sepulchre, N.E. Leonard, Autonomous rigid body attitude synchronization. Automatica **45**(2), 572–577 (2009)

Chapter 7
Distributed Robust Attitude Containment Control for Multiple Rigid Body Sysems

This chapter investigates the distributed robust finite-time attitude containment control for multiple rigid bodies with uncertainties including parametric uncertainties, external disturbances and actuator failures. Two novel types of distributed control laws are designed corresponding to two different cases respectively, and both of them can drive the orientations of the followers into the convex hull formed by the orientations of leaders in a finite time. Simulation results show the effectiveness of the proposed design.

This chapter is organized as follows. Section 7.1 presents the system model, problem formulation, and some useful lemmas. Our main results are given in Sect. 7.2, and the simulation results are shown in Sect. 7.3. The conclusion are located at the end of this chapter.

7.1 Problem Formulation

In this chapter, it is assumed that there are $m + n$ rigid bodies in the multiple rigid bodies system. m of them are considered as static leaders, i.e., the kth ($k \in \{1, \ldots, m\}$) leader rigid body's rotation remains constant represented as $q_k^0 \in R^3$, which is the Modified Rodriguez Parameter (MRP) of the kth leader, and its angular velocity remains zero. Besides, n of all the rigid bodies are considered as followers, and their attitude kinematic and dynamic equations are given by [1] as:

$$\frac{d}{dt} q_i = G(q_i)\omega_i \tag{7.1}$$

$$J_i \frac{d}{dt} \omega_i = (J_i \omega_i) \times \omega_i + \Gamma_i(t) u_i + \vartheta_i(t) \tag{7.2}$$

D. Yue et al., *Distributed Cooperative Control and Communication for Multi-agent Systems*, https://doi.org/10.1007/978-981-33-6718-0_7

where $q_i \in R^3$ is the MRP denoting the rotation from the body frame of the ith follower rigid body to the inertial frame, $\omega_i \in R^3$ denotes the angular velocity expressed in the body frame, the symbol '$\times$' denotes the cross product between two vectors, the moment of inertia matrix $J_i \in R^{3\times 3}$ is a constant positive definite matrix and $u_i \in R^3$ denotes the control torque of the ith rigid body. The time-varying diagonal matrix $\Gamma_i(t) \triangleq diag\{\Gamma_{i1}(t), \Gamma_{i2}(t), \Gamma_{i3}(t)\}$ with $0 < \Gamma_{ij}(t) \leq 1 (j = 1, 2, 3)$ is the actuation effectiveness matrix, and the time-varying vector $\vartheta_i(t) \in R^3$ represents the external disturbance. The function $G(q_i) = \frac{1}{2}(q_i^\times + q_i q_i^T + \frac{1-\|q_i\|^2}{2} I_3)$, where I_3 is the 3×3 identity matrix, $\| q_i \| \triangleq (q_i^T q_i)^{\frac{1}{2}}$ denotes the 2-norm of q_i, and $q_i^\times$ denotes the cross-product operator for the vector $q_i = \{q_{i1}, q_{i2}, q_{i3}\}^T \in R^3$ given by

$$q_i^\times = \begin{pmatrix} 0 & -q_{i3} & q_{i2} \\ q_{i3} & 0 & -q_{i1} \\ -q_{i2} & q_{i2} & 0 \end{pmatrix}$$

Define the matrix $B \triangleq diag\{\sum_{k=1}^m b_{1k}, \sum_{k=1}^m b_{2k}, \ldots, \sum_{k=1}^m b_{nk}\}$, where $b_{ik} = 1$ $(i = 1, \ldots, n.k = 1, \ldots, m)$ if the ith follower can receive the information of the kth leader, otherwise $b_{ik} = 0$.

Assumption 7.1 The time-varying actuation effectiveness matrix $\Gamma_i(t) = diag\{\Gamma_{i1}(t), \Gamma_{i2}(t), \Gamma_{i3}(t)\}$ in (7.2) satisfies the following inequality:

$$\Gamma_{ij}(t) \geq \gamma_i, \qquad j = 1, 2, 3$$

where γ_i is a positive constant.

Assumption 7.2 The time-varying external disturbance $\vartheta_i(t)$ in (7.2) is bounded such that $\| \vartheta_i(t) \| \leq \xi_i$, where ξ_i is a positive constant.

Remark 7.1 In the actuation effectiveness matrix $\Gamma_i(t) \triangleq diag\{\Gamma_{i1}(t), \Gamma_{i2}(t), \Gamma_{i3}(t)\}$, $\Gamma_{ij} = 1$ means that the jth actuator of the ith rigid body is healthy, and $0 < \Gamma_{ij} < 1$ implies that the jth actuator of the ith rigid body is partially failed. Assumption 7.1 implies that none of the actuators is totally failed.

Assumption 7.3 The graph $\mathcal{G}$ is connected, and at least one follower can receive the information of each leader, i.e., at least one b_{ik} equals to 1 for each k.

Lemma 7.1 *The matrix $(L + B)$ is positive definite under Assumption 7.3.*

The proof of Lemma 7.1 is similar with that of Lemma 3 in [2] and we omit it here.

Definition 7.1 ([3]) Let $X = \{x_1, x_2, \ldots, x_k\}$ be a set in a real vector space $V \subseteq R^p$. The convex hull $Co(X)$ of set X is defined as $Co(X) \triangleq \{\sum_{i=1}^k \alpha_i x_i | x_i \in X, \alpha_i \geq 0, \sum_{i=1}^k \alpha_i = 1, i = 1, 2, \ldots, k\}$.

The objective of this chapter is designing the attitude containment controller u_i for the follower rigid bodies, which can drive the followers' orientations $q_i (i = 1, \ldots, n)$ into the convex hull formed by the leaders' orientations, i.e. $Co(q_k^0, k = 1, 2, \ldots, m)$, in a finite time. The impacts of system parametric uncertainties, external disturbances and actuator failures will be considered, and the designed control torque u_i for the ith follower rigid body only contains the information of the followers in the neighboring set N_i and the leaders to which the ith follower rigid body can access.

Here we quote some lemmas which will be utilized in the following sections.

Lemma 7.2 ([4]) *Consider system $\dot{x} = f(x)$ where $f(0) = 0$ and $x \in R^n$. Suppose there exists a continuously differentiable, positive definite function $V(x) : U \to R$ defined on an open neighborhood U of the origin such that $\dot{V}(x) + c(V(x))^{\alpha} \leq 0$ on U for some $c > 0$ and $\alpha \in (0, 1)$. Then the origin is a finite-time stable equilibrium of system $\dot{x} = f(x)$ and the finite settling time T satisfies $T \leq \frac{V(x(0))^{1-\alpha}}{c(1-\alpha)}$ where x_0 is the initial value of the trajectory. If $U = R^n$ and V is radially unbounded, the origin is a globally finite-time stable equilibrium.*

Lemma 7.3 ([5]) *Consider the system $\dot{x} = f(x)$. Suppose that there exist continuously differentiable, positive definite function $V(x)$, scalars $\lambda > 0, 0 < \alpha < 1$ and $0 < \eta < \infty$ such that $\dot{V}(x) \leq -\lambda(V(x))^{\alpha} + \eta$. Then the trajectory of system $\dot{x} = f(x)$ converges to set $\{x : (V(x))^{\alpha} \leq \frac{\eta}{(1-\theta)\lambda}\}$ in a finite time and the finite settling time T satisfies $T \leq \frac{(V(x_0))^{1-\alpha}}{\lambda\theta(1-\alpha)}$ where x_0 is the initial value of the trajectory and $0 < \theta < 1$.*

Lemma 7.4 ([6]) *For $x_i \in R, i = 1, \ldots, n$, $0 < p \leqslant 1$, then $(|x_1| + \cdots |x_n|)^p \leq |x_1|^p + \cdots + |x_n|^p \leq n^{1-p}(|x_1| + \cdots |x_n|)^p$.*

7.2 Attitude Containment Control Law Design

In this section, control design for the attitude containment problem will be carried out by employing the sliding-mode control method. In Sect. 7.2.1, the sliding-mode surface is constructed such that the state of each follower on the surface will converge to the desired value in a finite time. In Sects. 7.2.2 and 7.2.3, under Assumptions 7.1 and 7.2 with that γ_i and ξ_i are known to be positive constants or unknown, distributed sliding-mode control or adaptive distributed sliding-mode control is designed respectively.

7.2.1 The Sliding-mode Surface Design

Construct the reference attitude signal q_{di} for each follower rigid body $i (i = 1, 2, \ldots n)$ as

$$q_{di} = \frac{\sum_{j=1}^{n} a_{ij}q_j + \sum_{k=1}^{m} b_{ik}q_k^0}{\sum_{j=1}^{n} a_{ij} + \sum_{k=1}^{m} b_{ik}} \tag{7.3}$$

Note that $\sum_{j=1}^{n} a_{ij} + \sum_{k=1}^{m} b_{ik}$ is strictly positive since the graph $\mathcal{G}$ is connected, and hence the reference attitude signal (7.3) is well defined. Moreover, for each follower agent i, the reference attitude signal q_{di} can be obtained in a distributed way since q_{di} only contains the information of the neighbors of agent i and the leaders to which agent i can access.

Defining the difference between q_i and q_{di} as $y_i \triangleq q_i - q_{di}$, the following lemma can be obtained.

Lemma 7.5 *If $y_i = 0$ for each follower agent i $(i = 1, 2, \ldots, n)$, then q_i is contained in the convex hull $Co(q_k^0, k = 1, \ldots, m)$ under Assumption 7.3, i.e. $q_i = \sum_{k=1}^{m} c_{ik}q_k^0$ where the constants c_{ik} $(i = 1, \ldots, n, k = 1, \ldots, m)$ satisfy $\sum_{k=1}^{m} c_{ik} = 1$ and $c_{ik} \geq 0$.*

Proof Defining the variable $e_i \triangleq q_i - \sum_{k=1}^{m} c_{ik}q_k^0$ where c_{ik} $(i = 1, 2, \ldots, n, k = 1, 2, \ldots, m)$ are some constants determined later, we can obtain

$$\begin{aligned} y_i &= \frac{(\sum_{j=1}^{n} a_{ij} + \sum_{k=1}^{m} b_{ik})q_i - (\sum_{j=1}^{n} a_{ij}q_j + \sum_{k=1}^{m} b_{ik}q_k^0)}{\sum_{j=1}^{n} a_{ij} + \sum_{k=1}^{m} b_{ik}} \\ &= \frac{\sum_{j=1}^{n} a_{ij}(q_i - q_j) + \sum_{k=1}^{m} b_{ik}(q_i - q_k^0)}{\sum_{j=1}^{n} a_{ij} + \sum_{k=1}^{m} b_{ik}} \\ &= \frac{\sum_{j=1}^{n} a_{ij}\big((e_i - e_j) + \sum_{k=1}^{m}(c_{ik} - c_{jk})q_k^0\big) + \sum_{k=1}^{m} b_{ik}(q_i - q_k^0)}{\sum_{j=1}^{n} a_{ij} + \sum_{k=1}^{m} b_{ik}} \\ &= \frac{\sum_{j=1}^{n} l_{ij}e_j + \sum_{j=1}^{n}\sum_{k=1}^{m} a_{ij}(c_{ik} - c_{jk})q_k^0 + \sum_{k=1}^{m} b_{ik}(q_i - q_k^0)}{\sum_{j=1}^{n} a_{ij} + \sum_{k=1}^{m} b_{ik}} \\ &= \frac{\sum_{j=1}^{n} l_{ij}e_j + (\sum_{k=1}^{m} b_{ik})q_i + \sum_{k=1}^{m}\big(\sum_{j=1}^{n} a_{ij}(c_{ik} - c_{jk}) - b_{ik}\big)q_k^0}{\sum_{j=1}^{n} a_{ij} + \sum_{k=1}^{m} b_{ik}} \end{aligned} \tag{7.4}$$

where l_{ij} is the element in Laplacian matrix L.

Next we will analyze the solution c_{ik} $(i = 1, 2, \ldots, n, k = 1, 2, \ldots, m)$ of the following equalities

$$\sum_{j=1}^{n} a_{ij}(c_{ik} - c_{jk}) - b_{ik} = -\left(\sum_{k=1}^{m} b_{ik}\right)c_{ik}, \qquad i = 1, \ldots, n, k = 1, \ldots, m \tag{7.5}$$

Obviously, (7.5) can be equivalently rewritten as

$$\sum_{j=1}^{n} l_{ij}c_{jk} + \left(\sum_{k=1}^{m} b_{ik}\right)c_{ik} = b_{ik} \tag{7.6}$$

which can also be expressed in a vector form as

$$(L + B)\mathbf{c}_k = \mathbf{b}_k, \qquad k = 1, \ldots, m \tag{7.7}$$

where $\mathbf{c}_k \triangleq (c_{1k}, \ldots, c_{nk})^T$ and $\mathbf{b}_k \triangleq (b_{1k}, \ldots, b_{nk})^T$. Equation (7.7) implies that the solution c_{ik} of (7.5) is existent and unique since $(L + B)$ is positive definite shown in Lemma 7.1. The following equality can be obtained from (7.7)

$$(L + B)\sum_{k=1}^{m} \mathbf{c}_k = \sum_{k=1}^{m} \mathbf{b}_k \tag{7.8}$$

Moreover

$$(L + B)\mathbf{1}^n = B\mathbf{1}^n = \left(\sum_{k=1}^{m} b_{1k}, \ldots, \sum_{k=1}^{m} b_{nk}\right)^T = \sum_{k=1}^{m} \mathbf{b}_k \tag{7.9}$$

From (7.8) and (7.9), we have $(L + B)\sum_{k=1}^{m} \mathbf{c}_k = (L + B)\mathbf{1}^n$, which implies that $\sum_{k=1}^{m} \mathbf{c}_k = \mathbf{1}^n$ since $(L + B)$ is positive definite. Therefore $\sum_{k=1}^{m} c_{ik} = 1$ for $i = 1, \ldots, n$.

Case A: For a selected k, assume that all elements in vector $\mathbf{c}_k$ are negative. From (7.6) and the fact that $\sum_{i=1}^{n} l_{ij} = 0$, we have

$$\begin{aligned}
\sum_{i=1}^{n} b_{ik} &= \sum_{i=1}^{n}\left[\sum_{j=1}^{n} l_{ij}c_{jk} + \left(\sum_{k=1}^{m} b_{ik}\right)c_{ik}\right] \\
&= \sum_{j=1}^{n}\left(\sum_{i=1}^{n} l_{ij}\right)c_{jk} + \sum_{i=1}^{n}\left(\sum_{k=1}^{m} b_{ik}\right)c_{ik} \\
&= \sum_{i=1}^{n}\left(\sum_{k=1}^{m} b_{ik}\right)c_{ik}
\end{aligned} \tag{7.10}$$

Assumption 7.3 implies that $\sum_{i=1}^{n} b_{ik} > 0$ for each k. However, the inequality $\sum_{i=1}^{n}(\sum_{k=1}^{m} b_{ik})c_{ik} \le 0$ is guaranteed since $c_{ik}(i = 1, \ldots n)$ are assumed to be negative in this case. The above two inequalities contradict (7.10). Therefore, the assumption that all elements in vector $\mathbf{c}_k$ are negative is false.

Case B: For a selected k, assume that there are $\nu(1 \le \nu < n)$ elements in vector $\mathbf{c}_k$ are negative and the rest are non-negative. Without loss of generality, we assume that the first ν elements in $\mathbf{c}_k$, i.e. $c_{1k}, c_{2k}, \ldots, c_{\nu k}$, are negative. Equation (7.6) and the property $\sum_{i=1}^{\nu} l_{ij} + \sum_{i=\nu+1}^{n} l_{ij} = \sum_{i=1}^{n} l_{ij} = 0$ imply

$$\begin{aligned}
\sum_{i=1}^{\nu} b_{ik} &= \sum_{i=1}^{\nu}\Big[\sum_{j=1}^{n} l_{ij}c_{jk} + (\sum_{k=1}^{m} b_{ik})c_{ik}\Big] \\
&= \sum_{i=1}^{\nu}\Big(\sum_{j=1}^{\nu} l_{ij}c_{jk} + \sum_{j=\nu+1}^{n} l_{ij}c_{jk}\Big) + \sum_{i=1}^{\nu}\Big(\sum_{k=1}^{m} b_{ik}\Big)c_{ik} \\
&= \sum_{j=1}^{\nu}\Big(\sum_{i=1}^{\nu} l_{ij}\Big)c_{jk} + \sum_{j=\nu+1}^{n}(\sum_{i=1}^{\nu} l_{ij})c_{jk} + \sum_{i=1}^{\nu}\Big(\sum_{k=1}^{m} b_{ik}\Big)c_{ik} \\
&= -\sum_{j=1}^{\nu}\Big(\sum_{i=\nu+1}^{n} l_{ij}\Big)c_{jk} + \sum_{j=\nu+1}^{n}\Big(\sum_{i=1}^{\nu} l_{ij}\Big)c_{jk} + \sum_{i=1}^{\nu}(\sum_{k=1}^{m} b_{ik})c_{ik} \\
&= -\sum_{i=1}^{\nu}\Big(\sum_{j=\nu+1}^{n} l_{ji}\Big)c_{ik} + \sum_{i=\nu+1}^{n}\Big(\sum_{j=1}^{\nu} l_{ji}\Big)c_{ik} + \sum_{i=1}^{\nu}\Big(\sum_{k=1}^{m} b_{ik}\Big)c_{ik}
\end{aligned} \tag{7.11}$$

Since $b_{ik} \geq 0, l_{ij} \leq 0$ for $i \neq j$, $c_{ik} < 0 (i = 1, \ldots, \nu)$ and $c_{jk} \geq 0 (j = \nu + 1, \ldots, n)$, equation (7.11) leads to

$$\sum_{i=1}^{\nu} b_{ik} \leq -\sum_{i=1}^{\nu}\Big(\sum_{j=\nu+1}^{n} l_{ji}\Big)c_{ik} \tag{7.12}$$

Moreover, we can obtain $-\sum_{i=1}^{\nu}(\sum_{j=\nu+1}^{n} l_{ji})c_{ik} \leq 0$, and the equality holds up if and only if $l_{ji} = 0$ for each $j = \nu + 1, \ldots, n$ and $i = 1, \ldots, \nu$, which implies that the graph $\mathcal{G}$ is not connected and contradicts Assumption 7.3. Therefore, the strict inequality $-\sum_{i=1}^{\nu}(\sum_{j=\nu+1}^{n} l_{ji})c_{ik} < 0$ is guaranteed. This together with the fact $\sum_{i=1}^{\nu} b_{ik} \geq 0$ contradicts (7.12). Therefore, the assumption that the first ν elements in $\mathbf{c}_k$ are negative is false.

From **Case A** and **Case B**, we can obtain that all the elements in $\mathbf{c}_k$ are non-negative.

Based on the above analyses, it can be concluded that the solution c_{ik} of (7.5) satisfies $c_{ik} \geq 0$ and $\sum_{k=1}^{m} c_{ik} = 1$ for $i = 1, \ldots, n, k = 1, \ldots, m$. Substituting the solution c_{ik} of (7.5) into (7.4) gives

$$\begin{aligned}
\Big(\sum_{j=1}^{n} a_{ij} + \sum_{k=1}^{m} b_{ik}\Big)y_i &= \sum_{j=1}^{n} l_{ij}e_j + \Big(\sum_{k=1}^{m} b_{ik}\Big)\Big(q_i - \sum_{k=1}^{m} c_{ik}q_k^0\Big) \\
&= \sum_{j=1}^{n} l_{ij}e_j + \Big(\sum_{k=1}^{m} b_{ik}\Big)e_i
\end{aligned}$$

which can be rewritten in an integrated form as

$$\big[(D + B) \otimes I_3\big]\mathbf{y} = \big[(L + B) \otimes I_3\big]\mathbf{e}$$

where $\otimes$ denotes the Kronecker product, I_3 is the three-order identity matrix, $\mathbf{y} \triangleq (y_1^T, \ldots, y_n^T)^T$ and $\mathbf{e} \triangleq (e_1^T, \ldots, e_n^T)^T$.

If $y_i = 0$ for each follower agent $i (i = 1, 2, \ldots n)$, then $\left[(D + B) \otimes I_3\right]\mathbf{y} = 0$, which implies $\left[(L + B) \otimes I_3\right]\mathbf{e} = 0$. Therefore we can obtain $\mathbf{e} = 0$ since the matrix $(L + B)$ is positive definite as shown in Lemma 7.1. This indicates $q_i = \sum_{k=1}^{m} c_{ik} q_k^0$ where $c_{ik} \geq 0$ and $\sum_{k=1}^{m} c_{ik} = 1$, which concludes the proof.

Lemma 7.5 states that if each q_i converges to q_{di}, the followers' orientations $q_i (i = 1, \ldots, n)$ will converge to $\sum_{k=1}^{m} c_{ik} q_k^0$, which is contained in the convex hull $Co(q_k^0, k = 1, \ldots, m)$ formed by the leaders' orientations.

Define the function

$$sig(y_i)^{\rho_i} \triangleq (|y_{i1}|^{\rho_i} sign(y_{i1}), |y_{i2}|^{\rho_i} sign(y_{i2}), |y_{i3}|^{\rho_i} sign(y_{i3}))^T$$

where y_{i1}, y_{i2} and y_{i3} are the three elements in vector $y_i \in R^3$ respectively, and $sign(\cdot)$ denotes the sign function.

Using the reference attitude signal q_{di} and the variable y_i, define the variable

$$\alpha_i \triangleq (G(q_i))^{-1} \left[\frac{d}{dt} q_{di} - k_i (1 + \parallel q_i \parallel^2) sig(y_i)^{\rho_i} \right], \qquad i = 1, \ldots, n \tag{7.13}$$

where $k_i > 0$, $0 < \rho_i < 1$ are constants.

Remark 7.2 From (7.1), it can be seen that

$$\frac{d}{dt} q_{di} = \frac{\sum_{j=1}^{n} a_{ij} G(q_j) \omega_j}{\sum_{j=1}^{n} a_{ij} + \sum_{k=1}^{m} b_{ik}}$$

This implies that the derivative of the reference attitude signal q_{di} with respect to time t can be easily computed by agent i based on the information obtained from its neighbors. Therefore, the variable α_i can be obtained by agent i through local information exchanges.

Introducing the following variable for each follower i

$$s_i \triangleq \omega_i - \alpha_i \tag{7.14}$$

the sliding-mode surface is constructed as

$$s_i = 0 \tag{7.15}$$

Lemma 7.6 *Under Assumption 7.3, consider the kinematic and dynamic systems (7.1)–(7.2) evolved on the sliding-mode surface (7.15) for each follower i, then each follower's orientation q_i converges to the convex hull formed by the leaders' orientations and its angular velocity ω_i converges to zero in a finite time, respectively.*

Proof Once on the sliding-mode surface $s_i = 0$, we have

$$\omega_i - \alpha_i = 0 \tag{7.16}$$

By (7.1), (7.13) and (7.16), we obtain

$$\begin{aligned}\frac{d}{dt}y_i &= \frac{d}{dt}q_i - \frac{d}{dt}q_{di}\\ &= G(q_i)\omega_i - \left[G(q_i)\alpha_i + k_i(1+ \| q_i \|^2)sig(y_i)^{\rho_i}\right]\\ &= -k_i(1+ \| q_i \|^2)sig(y_i)^{\rho_i}\end{aligned}$$

Consider the following Lyapunov function

$$V_{y_i} = \frac{1}{2}y_i^T y_i$$

By Lemma 7.4, the derivative of V_{y_i} with respect to time t satisfies

$$\begin{aligned}\frac{d}{dt}V_{y_i} &= y_i^T\left(\frac{d}{dt}y_i\right) = -k_i(1+ \| q_i \|^2)y_i^T sig(y_i)^{\rho_i} \leq -k_i y_i^T sig(y_i)^{\rho_i} \\ &= -k_i\sum_{j=1}^{3}|y_{ij}|^{\rho+1} \leq -k_i\left(\sum_{j=1}^{3}y_{ij}^2\right)^{\frac{\rho+1}{2}} \leq -2^{\frac{\rho+1}{2}}k_i V_{y_i}^{\frac{\rho+1}{2}}\end{aligned} \tag{7.17}$$

Based on the above inequality and Lemma 7.2, we obtain that y_i converges to 0 in a finite time, and the settling time is given by $T_i \leq \frac{(2V_{y_i}(y_i(0)))^{\frac{1-\rho}{2}}}{k_i(1-\rho)}$, where $V_{y_i}(y_i(0))$ is the initial value of $V_{y_i}(y_i)$. Therefore, all y_i for $i = 1, \dots, n$ converge to 0 in the finite time $T \triangleq max\{T_1, \dots, T_n\}$, which implies that each follower's orientation q_i converges to $\sum_{k=1}^{m} c_{ik}q_k^0$ with $\sum_{k=1}^{m} c_{ik} = 1$ and $c_{ik} \geqslant 0$ in the finite time T by Lemma 7.5. Moreover, since $\sum_{k=1}^{m} c_{ik}q_k^0$ is a constant, the angular velocity ω_i can achieve 0 in the finite time T based on (7.1).

Lemma 7.6 shows that once the orientation and angular velocity of each follower rigid body i is on the sliding-mode surface (7.15), all followers' orientations will converge to the convex hull formed by the leaders' orientations and their angular velocities will achieve zero in a finite time. Therefore, The distributed control law will be designed such that the orientation and angular velocity of each follower can be driven on the sliding-mode surface (7.15).

7.2.2 Sliding-mode Attitude Containment Control Design

This section considers the case in which the positive constants γ_i and ξ_i in Assumptions 7.1 and 7.2 are known for the ith follower rigid body, and the moment of inertia matrix J_i can be obtained accurately. The following result is concluded based on the Lyapunov method.

Theorem 7.1 *Under the Assumptions 7.1, 7.2 and 7.3 with the known positive constants γ_i and ξ_i, construct the following controller*

$$u_i = -k_{i1} sig(s_i)^{\rho_i} - \left(\| J_i \| \| \omega_i \|^2 + \| J_i \| \| \frac{d}{dt}\alpha_i \| + \xi_i \right) \frac{s_i}{\gamma_i \| s_i \|} \tag{7.18}$$

where α_i and s_i are defined in (7.13) and (7.14) respectively, and $k_{i1} > 0$ and $0 < \rho_i < 1$ are constants. Under the controller (7.18), the followers' orientations will converge to the convex hull formed by the leaders' orientations and the followers' angular velocities will converge to zero in a finite time, respectively.

Proof Consider the following Lyapunov function

$$\overline{V}_i = \frac{1}{2} s_i^T J_i s_i$$

Since the moment of inertia matrix J_i is positive definite, we obtain

$$\frac{1}{2}\lambda_{i,min} \| s_i \|^2 \leq \overline{V}_i \leq \frac{1}{2}\lambda_{i,max} \| s_i \|^2$$

where $\lambda_{i,min}$ and $\lambda_{i,max}$ are the minimum and maximum eigenvalues of J_i respectively. The derivative of $\overline{V}_i$ with respect to time t satisfies

$$\begin{aligned} \frac{d}{dt}\overline{V}_i &= s_i^T J_i \left(\frac{d}{dt}\omega_i - \frac{d}{dt}\alpha_i \right) \\ &= s_i^T \left[(J_i \omega_i) \times \omega_i + \Gamma_i u_i + \vartheta_i - J_i \frac{d}{dt}\alpha_i \right] \\ &\leq \left(\| J_i \| \| \omega_i \|^2 + \| J_i \| \| \frac{d}{dt}\alpha_i \| + \xi_i \right) \| s_i \| + s_i^T \Gamma_i u_i \end{aligned} \tag{7.19}$$

Since

$$s_i^T \Gamma_i s_i \geq \gamma_i \| s_i \|^2$$

and

$$s_i^T \Gamma_i sig(s_i)^{\rho_i} = \sum_{j=1}^{3} \Gamma_{ij} s_{ij} |s_{ij}|^{\rho_i} sign(s_{ij}) \geq \gamma_i \sum_{j=1}^{3} |s_{ij}|^{\rho_i + 1} \geq \gamma_i (\sum_{j=1}^{3} s_{ij}^2)^{\frac{\rho_i+1}{2}}$$

substituting (7.18) into (7.19) produces

$$\frac{d}{dt}\overline{V}_i \leq -k_{i1}\gamma_i\left(\sum_{j=1}^{3}s_{ij}^2\right)^{\frac{\rho_i+1}{2}} \leq -k_{i1}\gamma_i\left(\frac{2\overline{V}_i}{\lambda_{i,max}}\right)^{\frac{\rho_i+1}{2}} \tag{7.20}$$

where $s_{ij}(j = 1, 2, 3)$ are the three elements in vector s_i. By Lemma 7.2, equation (7.20) implies that $s_i(i = 1, \ldots, n)$ will achieve zero in a finite time, which ends the proof based on Lemma 7.6.

Remark 7.3 Similar to the controller in [7], the derivative of α_i with respect to time t is required in the controller (7.18). In other word, the angular accelerations of neighboring rigid bodies are necessary to implement the controller (7.18). This requirement increases the cost and complexity related to sensors and communication flow among neighboring rigid bodies. How to remove the item $\frac{d}{dt}\alpha_i$ in the controller remains an open problem, and will be studied in our future works.

Remark 7.4 Note that the variable α_i and $\frac{d}{dt}\alpha_i$ only contain the information of the neighbors of agent i and the leaders to which agent i can access, which implies that (7.18) is a distributed controller.

It is to be noted that the controller (7.18) is not continuous over the sliding-mode surface $s_i = 0$, which may lead to chattering. In order to avoid this undesirable phenomenon, the following continuous controller is designed based on (7.18).

Theorem 7.2 *Under the Assumptions 7.1, 7.2 and 7.3 with the known positive constants γ_i and ξ_i, construct the following controller*

$$u_i = \begin{cases} -k_{i1}sig(s_i)^{\rho_i} - (\| J_i \| \| \omega_i \|^2 + \| J_i \| \| \frac{d}{dt}\alpha_i \| + \xi_i)\frac{s_i}{\gamma_i \| s_i \|} \\ \quad if \quad (\| J_i \| \| \omega_i \|^2 + \| J_i \| \| \frac{d}{dt}\alpha_i \| + \xi_i) \| s_i \| > \delta_i, \\ -k_{i1}sig(s_i)^{\rho_i} - (\| J_i \| \| \omega_i \|^2 + \| J_i \| \| \frac{d}{dt}\alpha_i \| + \xi_i)^2 \frac{s_i}{\gamma_i \delta_i} \\ \quad if \quad (\| J_i \| \| \omega_i \|^2 + \| J_i \| \| \frac{d}{dt}\alpha_i \| + \xi_i) \| s_i \| \leq \delta_i \end{cases} \tag{7.21}$$

where α_i and s_i are defined in (7.13) and (7.14) respectively, and $k_{i1} > 0$, $0 < \rho_i < 1$, $\delta_i > 0$ are constants. Under the controller (7.21), s_i will be driven into a neighborhood of the sliding-mode surface (7.15) in a finite time. Furthermore, the followers' orientations will converge to an arbitrary small neighborhood of the convex hull formed by the leaders' orientations in a finite time.

Proof Consider the same Lyapunov function $\overline{V}_i$ as that in the proof of Theorem 7.1. If $(\| J_i \| \| \omega_i \|^2 + \| J_i \| \| \frac{d}{dt}\alpha_i \| + \xi_i) \| s_i \| > \delta_i$, by a similar proof of Theorem 7.1, we have

$$\frac{d}{dt}\overline{V}_i \leq -k_{i1}\gamma_i\left(\frac{2\overline{V}_i}{\lambda_{i,max}}\right)^{\frac{\rho_i+1}{2}}$$

If $(\| J_i \| \| \omega_i \|^2 + \| J_i \| \| \frac{d}{dt}\alpha_i \| + \xi_i) \| s_i \| \leq \delta_i$, we have

$$\frac{d}{dt}\overline{V}_i \leq -k_{i1}\gamma_i\left(\frac{2\overline{V}_i}{\lambda_{i,max}}\right)^{\frac{\rho_i+1}{2}} + \left(1 - (\| J_i \|\| \omega_i \|^2 + \| J_i \|\| \frac{d}{dt}\alpha_i \| + \xi_i\right)\frac{\| s_i \|}{\delta_i}\Bigg)$$
$$\cdot\left(\| J_i \|\| \omega_i \|^2 + \| J_i \|\| \frac{d}{dt}\alpha_i \| + \xi_i\right) \| s_i \|$$
$$\leq -k_{i1}\gamma_i\left(\frac{2\overline{V}_i}{\lambda_{i,max}}\right)^{\frac{\rho_i+1}{2}} + \delta_i$$

Based on the above analyses, we obtain

$$\frac{d}{dt}\overline{V}_i \leq -k_{i1}\gamma_i\left(\frac{2\overline{V}_i}{\lambda_{i,max}}\right)^{\frac{\rho_i+1}{2}} + \delta_i \tag{7.22}$$

By Lemma 7.3, inequality (7.22) implies s_i will converge to following set in a finite time as

$$\widetilde{S}_i \triangleq \left\{ s_i : (\overline{V}_i(s_i))^{\frac{\rho_i+1}{2}} \leq \left(\frac{\lambda_{i,max}}{2}\right)^{\frac{\rho_i+1}{2}} \frac{\delta_i}{k_{i1}\gamma_i(1-\theta_i)} \right\}$$

and the settling time $T_i \leq (\frac{\lambda_{i,max}}{2})^{\frac{\rho_i+1}{2}} \frac{2(\overline{V}_i(s_i(0)))^{\frac{1-\rho_i}{2}}}{k_{i1}\gamma_i\theta_i(1-\rho_i)}$ where $\overline{V}_i(s_i(0))$ is the initial value of $\overline{V}_i(s_i)$ and $0 < \theta_i < 1$. Since $(\overline{V}_i(s_i))^{\frac{\rho_i+1}{2}} = (\frac{1}{2}s_i^T J_i s_i)^{\frac{\rho_i+1}{2}} \geq (\frac{\lambda_{i,min}}{2})^{\frac{\rho_i+1}{2}} \| s_i \|^{\rho_i+1}$, we then obtain that s_i will converge to the following set in the finite time T_i as

$$\overline{\widetilde{S}}_i \triangleq \left\{ s_i : \| s_i \| \leq \left(\frac{\lambda_{i,max}}{\lambda_{i,min}}\right)^{\frac{1}{2}} \left(\frac{\delta_i}{k_{i1}\gamma_i(1-\theta_i)}\right)^{\frac{1}{\rho_i+1}} \right\} \tag{7.23}$$

which is a neighborhood of the sliding-mode surface $s_i = 0$.

Next we prove that the followers' orientations will converge to an arbitrary small neighborhood of the convex hull formed by the leaders' orientations in a finite time by a similar method in [8]. Defining the constant $\overline{C}_i \triangleq (\frac{\lambda_{i,max}}{\lambda_{i,min}})^{\frac{1}{2}} (\frac{\delta_i}{k_{i1}\gamma_i(1-\theta_i)})^{\frac{1}{\rho_i+1}}$, we can obtain $|\{s_i\}_j| \leq \overline{C}_i (j = 1, 2, 3)$ since $\| s_i \| \leq \overline{C}_i$ as shown in (7.23), where the symbol $\{a\}_j$ denotes the jth element of vector a. Define the variable

$$\phi_{ij} \triangleq \{\omega_i\}_j - \left\{ G(q_i)^{-1} \frac{d}{dt} q_{di} \right\}_j + k_i\{(1+ \| q_i \|^2) G(q_i)^{-1} sig(y_i)^{\rho_i}\}_j \tag{7.24}$$

By the definition of s_i, we have $|\phi_{ij}| \leq \overline{C}_i$. Equation (7.24) implies

$$\{\omega_i\}_j - \left\{G(q_i)^{-1}\frac{d}{dt}q_{di}\right\}_j + \left(k_i - \frac{\phi_{ij}}{\{(1+\parallel q_i \parallel^2)G(q_i)^{-1}sig(y_i)^{\rho_i}\}_j}\right)$$
$$\cdot\{(1+\parallel q_i \parallel^2)G(q_i)^{-1}sig(y_i)^{\rho_i}\}_j = 0, \quad (7.25)$$

When $k_i - \frac{\phi_{ij}}{\{(1+\|q_i\|^2)G(q_i)^{-1}sig(y_i)^{\rho_i}\}_j} > 0$, Eq. (7.25) is still kept in the form of (7.15), which means that $\{(1+\parallel q_i \parallel^2)G(q_i)^{-1}sig(y_i)^{\rho_i}\}_j$ will converge to the following region in a finite time as

$$|\{(1+\parallel q_i \parallel^2)G(q_i)^{-1}sig(y_i)^{\rho_i}\}_j| \leq \frac{\overline{C}_i}{k_i} \quad (7.26)$$

From (7.26), we have

$$(1+\parallel q_i \parallel^2) \parallel G(q_i)^{-1}sig(y_i)^{\rho_i} \parallel \leq \sqrt{3}\frac{\overline{C}_i}{k_i} \quad (7.27)$$

Since $(G(q_i)^{-1})^T G(q_i)^{-1} = (G(q_i)G^T(q_i))^{-1} = (\frac{4}{1+\|q_i\|^2})^2 I_3$ as shown in [9] and $(sig(y_i)^{\rho_i})^T(sig(y_i)^{\rho_i}) = \sum_{j=1}^{3} |\{y_i\}_j|^{2\rho_i} \geq (\sum_{j=1}^{3}\{y_i\}_j^2)^{\rho_i} = \parallel y_i \parallel^{2\rho_i}$, we can obtain

$$\begin{aligned}\parallel G(q_i)^{-1}sig(y_i)^{\rho_i} \parallel &= \left((sig(y_i)^{\rho_i})^T(G(q_i)^{-1})^T G(q_i)^{-1}sig(y_i)^{\rho_i}\right)^{\frac{1}{2}} \\ &\geq \frac{4}{1+\parallel q_i \parallel^2} \parallel y_i \parallel^{\rho_i} \end{aligned} \quad (7.28)$$

Combining (7.27) and (7.28) gives that y_i will converge to the arbitrary small region

$$\left\{y_i : \parallel y_i \parallel \leq \left(\frac{\sqrt{3}}{4}\frac{\overline{C}_i}{k_i}\right)^{\frac{1}{\rho_i}}\right\}$$

in a finite time, which indicates that the followers' orientations will converge to an arbitrary small neighborhood of the convex hull formed by the leaders' orientations in a finite time.

Remark 7.5 From the expression (7.23), we can obtain that s_i will reach the sliding-mode surface $s_i = 0$ as close as possible by suitable selection of the parameters in the controller (7.21). Moreover, the neighborhood of the convex hull formed by the leaders' orientations can be arbitrary small by parametric selection in the controller based on the definition of $\overline{C}_i$.

7.2.3 Adaptive Sliding-mode Attitude Containment Control Design

In order to implement the controller (7.18) or (7.21), the upper bound of external disturbance ξ_i, lower bound of actuator effectiveness γ_i and moment of inertial matrix J_i are required. However, the derivation of these values is difficult and even impossible in many real systems. In this section, the distributed adaptive sliding-mode controller in the case of unknown parameters ξ_i, γ_i and J_i will be designed, which to make the control law more valuable in practical applications.

Theorem 7.3 *Under the Assumptions 7.1, 7.2 and 7.3 with the unknown positive constants γ_i and ξ_i, construct the controller*

$$u_i = -k_{i2} sig(s_i)^{\rho_i} - \left(k_{i3} + \frac{1}{2\varepsilon_i}\hat{\eta}_i \theta_i\right) s_i \tag{7.29}$$

and the adaptive law with the initial value $\hat{\eta}_i(0) > 0$

$$\frac{d}{dt}\hat{\eta}_i = \frac{\theta_i}{2\varepsilon_i \beta_i} \parallel s_i \parallel^2 - a_i \hat{\eta}_i \tag{7.30}$$

where $k_{i2} > 0$, $k_{i3} > 0$, $\varepsilon_i > 0$, $a_i > 0$, $0 < \rho_i < 1$ are some constants, the variable s_i is defined in (7.14), the variable θ_i is defined as

$$\theta_i \triangleq \parallel \omega_i \parallel^4 + \parallel \frac{d}{dt}\alpha_i \parallel + 1 \tag{7.31}$$

and the parameter β_i satisfies

$$\beta_i \geq \left(\frac{k_{i2}}{2a_i \widetilde{\alpha}_i}\right)^{\frac{2}{1-\rho_i}} \left(\frac{2}{\lambda_{i,max}}\right)^{\frac{1+\rho_i}{1-\rho_i}} \tag{7.32}$$

where $0 < \widetilde{\alpha}_i < 1$ and $\lambda_{i,max}$ is the maximum eigenvalue of J_i.

Then, s_i will be driven into a neighborhood of the sliding-mode surface (7.15) in a finite time. Furthermore, the followers' orientations will converge to an arbitrary small neighborhood of the convex hull formed by the leaders' orientations in a finite time.

Proof Consider the function

$$V_{0i} = \frac{1}{2} s_i^T J_i s_i$$

It is known from the positive definiteness of matrix J_i that there exist positive constants c_{i1} and c_{i2} such that $\parallel (J_i \omega_i) \times \omega_i \parallel \leq c_{i1} \parallel \omega_i \parallel^2$ and $\parallel J_i \frac{d}{dt}\alpha_i \parallel \leq c_{i2} \parallel \frac{d}{dt}\alpha_i \parallel$, then we have

$$
\begin{aligned}
\frac{d}{dt}V_{0i} &= s_i^T J_i\left(\frac{d}{dt}\omega_i - \frac{d}{dt}\alpha_i\right) \\
&= s_i^T\left[(J_i\omega_i)\times\omega_i + \Gamma_i u_i + \vartheta_i - J_i\frac{d}{dt}\alpha_i\right] \\
&\le c_{i1} \parallel s_i \parallel \parallel \omega_i \parallel^2 + c_{i2} \parallel s_i \parallel \parallel \frac{d}{dt}\alpha_i \parallel + \xi_i \parallel s_i \parallel + s_i^T\Gamma_i u_i \\
&\le \frac{1}{2\varepsilon_i}\left(c_{i1}^2 \parallel s_i \parallel^2 \parallel \omega_i \parallel^4 + c_{i2}^2 \parallel s_i \parallel^2 \parallel \frac{d}{dt}\alpha_i \parallel^2 + \xi_i^2 \parallel s_i \parallel^2\right) + \frac{3}{2}\varepsilon_i + s_i^T\Gamma_i u_i \\
&\le \frac{1}{2\varepsilon_i}\eta_i\theta_i \parallel s_i \parallel^2 + \frac{3}{2}\varepsilon_i + s_i^T\Gamma_i u_i
\end{aligned} \tag{7.33}
$$

where $\eta_i \triangleq max\{c_{i1}^2, c_{i2}^2, \xi_i^2\}$ is an unknown constant.

Defining the variable $\widetilde{\eta}_i \triangleq \eta_i - \gamma_i\hat{\eta}_i$ and the function $V_{\eta i} \triangleq \frac{\beta}{2\gamma_i}\widetilde{\eta}_i^2$, construct the following Lyapunov function

$$
V_i = V_{0i} + V_{\eta i}
$$

Using (7.29) and (7.33), the time derivative of V_i results in

$$
\begin{aligned}
\frac{d}{dt}V_i &= \frac{d}{dt}V_{0i} + \frac{d}{dt}V_{\eta i} \\
&\le \frac{1}{2\varepsilon_i}\eta_i\theta_i \parallel s_i \parallel^2 + \frac{3}{2}\varepsilon_i + s_i^T\Gamma_i\left[-k_{i2}sig(s_i)^{\rho_i} - \left(k_{i3} + \frac{1}{2\varepsilon_i}\hat{\eta}_i\theta_i\right)s_i\right] + \frac{\beta_i}{\gamma_i}\widetilde{\eta}_i\frac{d}{dt}\widetilde{\eta}_i \\
&= \frac{1}{2\varepsilon_i}\eta_i\theta_i \parallel s_i \parallel^2 + \frac{3}{2}\varepsilon_i - k_{i2}s_i^T\Gamma_i sig(s_i)^{\rho_i} - s_i^T\Gamma_i(k_{i3} + \frac{1}{2\varepsilon_i}\hat{\eta}_i\theta_i)s_i - \beta_i\widetilde{\eta}_i\frac{d}{dt}\hat{\eta}_i
\end{aligned} \tag{7.34}
$$

Let $s_{ij}(j = 1, 2, 3)$ are the three elements in vector s_i, we have

$$
\begin{aligned}
k_{i2}s_i^T\Gamma_i sig(s_i)^{\rho_i} &= k_{i2}\sum_{j=1}^{3}\Gamma_{ij}s_{ij}|s_{ij}|^{\rho_i}sign(s_{ij}) \\
&\ge k_{i2}\gamma_i\sum_{j=1}^{3}|s_{ij}|^{\rho_i+1} \ge k_{i2}\gamma_i(\sum_{j=1}^{3}s_{ij}^2)^{\frac{\rho_i+1}{2}} \ge k_{i2}\gamma_i\left(\frac{2V_{0i}}{\lambda_{i,max}}\right)^{\frac{\rho_i+1}{2}}
\end{aligned} \tag{7.35}
$$

Note that $\hat{\eta}_i(t) \ge 0$ since the the initial value $\hat{\eta}_i(0) > 0$ in adaptive law (7.30), we obtain

$$
s_i^T\Gamma_i\left(k_{i3} + \frac{1}{2\varepsilon_i}\hat{\eta}_i\theta_i\right)s_i \ge \gamma_i\left(k_{i3} + \frac{1}{2\varepsilon_i}\hat{\eta}_i\theta_i\right) \parallel s_i \parallel^2 \tag{7.36}
$$

Substituting (7.35) and (7.36) into (7.34) yields

$$
\begin{aligned}
\frac{d}{dt}V_i &\le -\gamma_i k_{i3} \parallel s_i \parallel^2 - k_{i2}\gamma_i\left(\frac{2}{\lambda_{i,max}}\right)^{\frac{\rho_i+1}{2}} (V_{0i})^{\frac{\rho_i+1}{2}} + \frac{3}{2}\varepsilon_i + a_i\beta_i\widetilde{\eta}_i\hat{\eta}_i \\
&\le -k_{i2}\gamma_i\left(\frac{2}{\lambda_{i,max}}\right)^{\frac{\rho_i+1}{2}}\left(V_{0i}^{\frac{\rho_i+1}{2}} + V_{\eta i}^{\frac{\rho_i+1}{2}}\right) \\
&\quad + k_{i2}\gamma_i\left(\frac{2}{\lambda_{i,max}}\right)^{\frac{\rho_i+1}{2}} V_{\eta i}^{\frac{\rho_i+1}{2}} + a_i\beta_i\widetilde{\eta}_i\hat{\eta}_i + \frac{3}{2}\varepsilon_i
\end{aligned} \tag{7.37}
$$

If $\frac{1}{2\gamma_i}\widetilde{\eta}_i^2 > 1$, we can get

$$
\begin{aligned}
k_{i2}\gamma_i\left(\frac{2}{\lambda_{i,max}}\right)^{\frac{\rho_i+1}{2}} V_{\eta i}^{\frac{\rho_i+1}{2}} &= k_{i2}\gamma_i\left(\frac{2\beta_i}{\lambda_{i,max}}\right)^{\frac{\rho_i+1}{2}}\left(\frac{1}{2\gamma_i}\widetilde{\eta}_i^2\right)^{\frac{\rho_i+1}{2}} \\
\le k_{i2}\gamma_i\left(\frac{2\beta_i}{\lambda_{i,max}}\right)^{\frac{\rho_i+1}{2}} \frac{1}{2\gamma_i}\widetilde{\eta}_i^2 &\le k_{i2}\left(\frac{2\beta_i}{\lambda_{i,max}}\right)^{\frac{\rho_i+1}{2}} \widetilde{\eta}_i^2
\end{aligned} \tag{7.38}
$$

Using (7.32) and (7.38), the following inequality can be obtained

$$
\begin{aligned}
& k_{i2}\gamma_i\left(\frac{2}{\lambda_{i,max}}\right)^{\frac{\rho_i+1}{2}} V_{\eta i}^{\frac{\rho_i+1}{2}} + a_i\beta_i\widetilde{\eta}_i\hat{\eta}_i \\
&\le k_{i2}\left(\frac{2\beta_i}{\lambda_{i,max}}\right)^{\frac{\rho_i+1}{2}} \widetilde{\eta}_i^2 + \frac{a_i\beta_i}{\gamma_i}\widetilde{\eta}_i(\eta_i - \widetilde{\eta}_i) \\
&= \left[k_{i2}\left(\frac{2\beta_i}{\lambda_{i,max}}\right)^{\frac{\rho_i+1}{2}} - \frac{a_i\beta_i}{\gamma_i}\right]\widetilde{\eta}_i^2 + \frac{a_i\beta_i}{\gamma_i}\eta_i\widetilde{\eta}_i \\
&\le \left[k_{i2}\left(\frac{2\beta_i}{\lambda_{i,max}}\right)^{\frac{\rho_i+1}{2}} - \frac{a_i\beta_i}{\gamma_i}\right]\widetilde{\eta}_i^2 + \frac{a_i\beta_i}{\gamma_i}\left[\left(1 - \widetilde{\alpha}_i\gamma_i\right)\widetilde{\eta}_i^2 + \frac{1}{4(1-\widetilde{\alpha}_i\gamma_i)}\eta_i^2\right] \\
&= \left[k_{i2}\left(\frac{2\beta_i}{\lambda_{i,max}}\right)^{\frac{\rho_i+1}{2}} - a_i\beta_i\widetilde{\alpha}_i\right]\widetilde{\eta}_i^2 + \frac{a_i\beta_i}{4\gamma_i(1-\widetilde{\alpha}_i\gamma_i)}\eta_i^2 \\
&\le \frac{a_i\beta_i}{4\gamma_i(1-\widetilde{\alpha}_i\gamma_i)}\eta_i^2
\end{aligned} \tag{7.39}
$$

If $\frac{1}{2\gamma_i}\widetilde{\eta}_i^2 \le 1$, we can get

$$
k_{i2}\gamma_i\left(\frac{2}{\lambda_{i,max}}\right)^{\frac{\rho_i+1}{2}} V_{\eta i}^{\frac{\rho_i+1}{2}} = k_{i2}\gamma_i\left(\frac{2\beta_i}{\lambda_{i,max}}\right)^{\frac{\rho_i+1}{2}}\left(\frac{1}{2\gamma_i}\widetilde{\eta}_i^2\right)^{\frac{\rho_i+1}{2}} \le k_{i2}\gamma_i\left(\frac{2\beta_i}{\lambda_{i,max}}\right)^{\frac{\rho_i+1}{2}}
$$

which further implies

$$
\begin{aligned}
& k_{i2}\gamma_i\left(\frac{2}{\lambda_{i,max}}\right)^{\frac{\rho_i+1}{2}} V_{\eta i}^{\frac{\rho_i+1}{2}} + a_i\beta_i\widetilde{\eta}_i\hat{\eta}_i \\
\leq\; & k_{i2}\gamma_i\left(\frac{2\beta_i}{\lambda_{i,max}}\right)^{\frac{\rho_i+1}{2}} + \frac{a_i\beta_i}{\gamma_i}\widetilde{\eta}_i(\eta_i - \widetilde{\eta}_i) \\
\leq\; & k_{i2}\gamma_i\left(\frac{2\beta_i}{\lambda_{i,max}}\right)^{\frac{\rho_i+1}{2}} + \frac{a_i\beta_i}{4\gamma_i}\eta_i^2
\end{aligned}
\tag{7.40}
$$

where the last inequality can be obtained since $\widetilde{\eta}_i(\eta_i - \widetilde{\eta}_i) = \widetilde{\eta}_i\eta_i - \widetilde{\eta}_i^2 \leq \frac{1}{4}\eta_i^2$. Define the constant

$$
C_i \triangleq max\left\{\frac{a_i\beta_i}{4\gamma_i(1-\widetilde{\alpha}_i\gamma_i)}\eta_i^2,\;\; k_{i2}\gamma_i\left(\frac{2\beta_i}{\lambda_{i,max}}\right)^{\frac{\rho_i+1}{2}} + \frac{a_i\beta_i}{4\gamma_i}\eta_i^2\right\}
$$

From (7.37), (7.39) and (7.40), the time derivative of Lyapunov function V_i becomes

$$
\begin{aligned}
\frac{d}{dt}V_i &\leq -k_{i2}\gamma_i\left(\frac{2}{\lambda_{i,max}}\right)^{\frac{\rho_i+1}{2}}\left(V_{0i}^{\frac{\rho_i+1}{2}} + V_{\eta i}^{\frac{\rho_i+1}{2}}\right) + C_i + \frac{3}{2}\varepsilon_i \\
&\leq -k_{i2}\gamma_i\left(\frac{2}{\lambda_{i,max}}\right)^{\frac{\rho_i+1}{2}} V_i^{\frac{\rho_i+1}{2}} + C_i + \frac{3}{2}\varepsilon_i
\end{aligned}
\tag{7.41}
$$

By Lemma 7.3, inequality (7.41) implies s_i will converge to the following set in a finite time

$$
S_i \triangleq \left\{s_i : (V_i(s_i,\widetilde{\eta}_i))^{\frac{\rho_i+1}{2}} \leq \frac{C_i + \frac{3}{2}\varepsilon_i}{(1-d)k_{i2}\gamma_i(\frac{2}{\lambda_{i,max}})^{\frac{\rho_i+1}{2}}}\right\}
$$

where $0 < d < 1$ is a constant, and the settling time $T_i \leq \frac{2(V_i(0))^{\frac{1-\rho_i}{2}}}{(1-\rho_i)dk_{i2}\gamma_i}(\frac{\lambda_{i,max}}{2})^{\frac{1+\rho_i}{2}}$ where $V_i(0)$ is the initial value of Lyapunov function $V_i(t)$. Since $V_i^{\frac{\rho_i+1}{2}} \geq V_{0i}^{\frac{\rho_i+1}{2}} \geq (\frac{\lambda_{i,min}}{2} \| s_i \|^2)^{\frac{\rho_i+1}{2}} \geq (\frac{\lambda_{i,min}}{2})^{\frac{\rho_i+1}{2}} \| s_i \|^{\rho_i+1}$ where $\lambda_{i,min}$ is the minimum eigenvalue of matrix J_i, we then obtain that s_i will converge to the following set in the finite time T_i as

$$
\overline{S}_i \triangleq \left\{s_i : \| s_i \| \leq \left(\frac{\lambda_{i,max}}{\lambda_{i,min}}\right)^{\frac{1}{2}}\left[\frac{C_i + \frac{3}{2}\varepsilon_i}{(1-d)k_{i2}\gamma_i}\right]^{\frac{1}{\rho_i+1}}\right\}
\tag{7.42}
$$

which is a neighborhood of the sliding-mode surface $s_i = 0$. The remaining proof is similar to that in Theorem 7.2 and we omit it here.

Remark 7.6 Selecting the parameter $\beta_i = (\frac{k_{i2}}{2a_i\widetilde{\alpha}_i})^{\frac{2}{1-\rho_i}}(\frac{2}{\lambda_{i,max}})^{\frac{1+\rho_i}{1-\rho_i}}$ in (7.30), the set $\overline{S}_i$ described in (7.42) can be expressed as

$$\overline{S}_i = \left\{ s_i : \| s_i \| \le max\{C_{1i}, C_{2i}\} \right\}$$

where

$$C_{1i} = \left(\frac{\lambda_{i,max}}{\lambda_{i,min}}\right)^{\frac{1}{2}} \left[\frac{k_{i2}^{\frac{1+\rho_i}{1-\rho_i}} \eta_i^2}{4\gamma_i^2(1-d)(1-\widetilde{\alpha}_i\gamma_i)a_i^{\frac{1+\rho_i}{1-\rho_i}}} \left(\frac{1}{2\widetilde{\alpha}_i}\right)^{\frac{2}{1-\rho_i}} \left(\frac{2}{\lambda_{i,max}}\right)^{\frac{1+\rho_i}{1-\rho_i}} + \frac{3}{2} \frac{\varepsilon_i}{(1-d)k_{i2}\gamma_i} \right]^{\frac{1}{1+\rho_i}}$$

$$C_{2i} = \left(\frac{\lambda_{i,max}}{\lambda_{i,min}}\right)^{\frac{1}{2}} \left[\frac{k_{i2}^{\frac{1+\rho_i}{1-\rho_i}} \eta_i^2}{4\gamma_i^2(1-d)a_i^{\frac{1+\rho_i}{1-\rho_i}}} \left(\frac{1}{2\widetilde{\alpha}_i}\right)^{\frac{2}{1-\rho_i}} \left(\frac{2}{\lambda_{i,max}}\right)^{\frac{1+\rho_i}{1-\rho_i}} + \frac{3}{2} \frac{\varepsilon_i}{(1-d)k_{i2}\gamma_i} \right.$$
$$\left. + \frac{1}{1-d} \left(\frac{k_{i2}}{\lambda_{i,max} a_i \widetilde{\alpha}_i}\right)^{\frac{1+\rho_i}{1-\rho_i}} \right]^{\frac{1}{1+\rho_i}}$$

From the expressions of C_{1i} and C_{2i}, it can be seen that s_i will reach the sliding-mode surface as close as possible by suitable selecting the parameters in the controller (7.29)–(7.30). Moreover, the neighborhood of the convex hull formed by the leaders' orientations can be arbitrary small by parametric selection in the controller.

Remark 7.7 Although the upper bound of external disturbance ξ_i and the lower bound of actuator effectiveness γ_i defined in Assumptions 7.1 and 7.2 are used in the preceding stability analysis, they are not required in the proposed controller (7.29)–(7.30). Moreover, the moment of inertia matrix J_i, which is difficult to measure in practical applications, is also not needed to implement the designed controller. Although the selection of parameter β_i is related to the maximum eigenvalue of J_i, we can choose β_i large enough to ensure the stability without the exact knowledge of J_i in practical applications.

7.3 Simulation Examples

This section presents the simulation results demonstrating the effectiveness of the controllers designed in Sect. 7.3. The results under the controllers (7.21) and (7.29)–(7.30) are given in Sects. 7.3.1 and 7.3.2 respectively.

Consider six rigid bodies with the communication topology shown in Fig. 7.1. The agents v_1, v_2, v_3 and v_4 are the followers whose attitude kinematic and dynamic equations are described as (7.1)–(7.2), and v_1^l and v_2^l are the static leaders. Obviously, the information flow between the four followers is presented by an undirected fix connected graph, and there is one follower can access to each leader. For simplicity and without loss of generality, the moment of inertia matrix of each rigid body is set as the identity matrix.

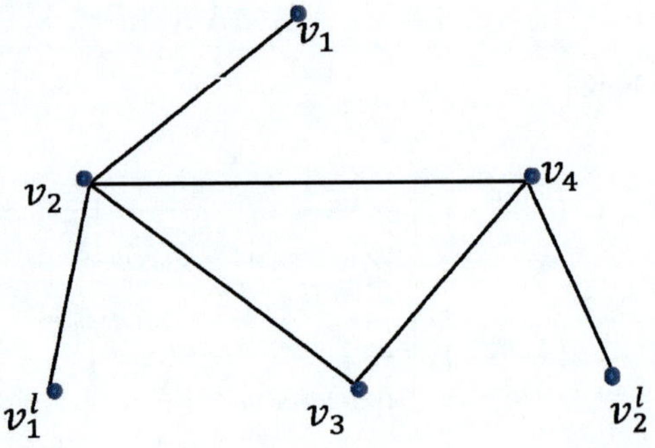

Fig. 7.1 The communication topology of six rigid bodies

7.3.1 Sliding-mode Attitude Containment Control

This section shows the simulation results under the sliding-mode attitude containment controller (7.21). The initial values of the followers' orientations are set as

$$q_1(0) = (-1.35, -1.75, -0.36)^T, \quad q_2(0) = (-0.63, 0.44, -1.50)^T$$
$$q_3(0) = (-0.21, -1.51, 1.81)^T, \quad q_4(0) = (-0.12, 1.23, 0.49)^T$$

and the initial values of the followers' angular velocities are

$$\omega_1(0) = (0.91, -1.15, 1.26)^T, \quad \omega_2(0) = (1.45, -2.08, -0.18)^T$$
$$\omega_3(0) = (-0.65, 0.16, -0.85)^T, \quad \omega_4(0) = (0.28, 0.59, 0.24)^T$$

The static leaders' orientations are considered to be

$$q_1^0 = (0, -0.2, 1)^T, \quad q_2^0 = (0.2, 0, 0.8)^T$$

Moreover, the external disturbance for each follower is

$$\vartheta_i(t) = 0.3(sin(t), cos(t), sin(2t))^T, \qquad i = 1, 2, 3, 4$$

and the actuation effectiveness matrix is assumed to be

$$\Gamma_i(t) = \begin{pmatrix} 0.75 + 0.25cos(t) & 0 & 0 \\ 0 & 0.6 + 0.4sin(2t) & 0 \\ 0 & 0 & 0.55 + 0.45cos(3t) \end{pmatrix}, i = 1, 2, 3, 4$$

Obviously, the bounds in Assumptions 7.1 and 7.2 are

$$\gamma_i = 0.1, \qquad \xi_i = 0.3$$

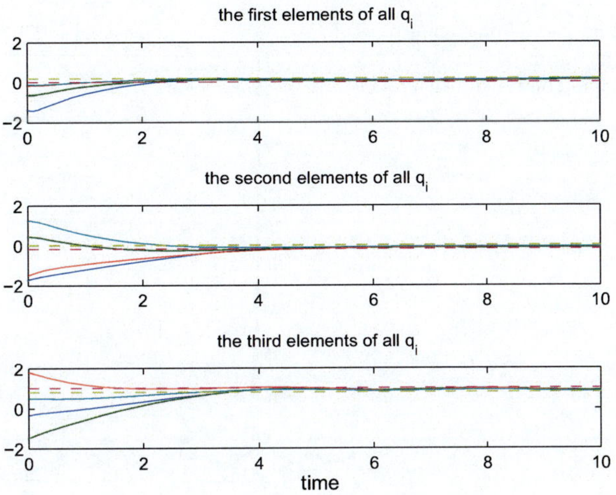

Fig. 7.2 Time responses of all rigid bodies' orientations under controller (7.21)

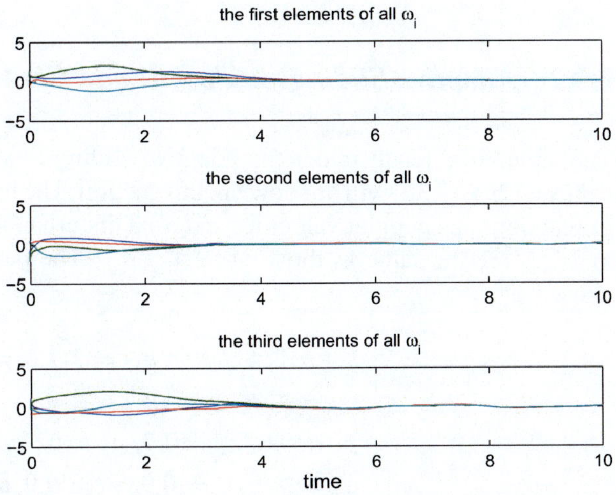

Fig. 7.3 Time responses of all rigid bodies' angular velocities under controller (7.21)

The parameters in the four follower' controllers (7.21) are set as $k_{11} = 10, k_{21} = 7, k_{31} = 10, k_{41} = 12, \rho_1 = 0.6, \rho_2 = 0.7, \rho_3 = 0.9, \rho_4 = 0.8, \delta_i = 10$ and $k_i = 0.7$ for $i = 1, 2, 3, 4$. The simulation results are shown in Figs. 7.2, 7.3 and 7.4. In Figs. 7.2 and 7.3, the dash lines and solid lines denote the time responses of the leaders' states and the followers' states respectively. It can be seen that, the four followers' orientations converge to the convex hull of the leaders' orientations in less than 5 s. Figure 7.4 depicts the norms of the input control signals of all four followers.

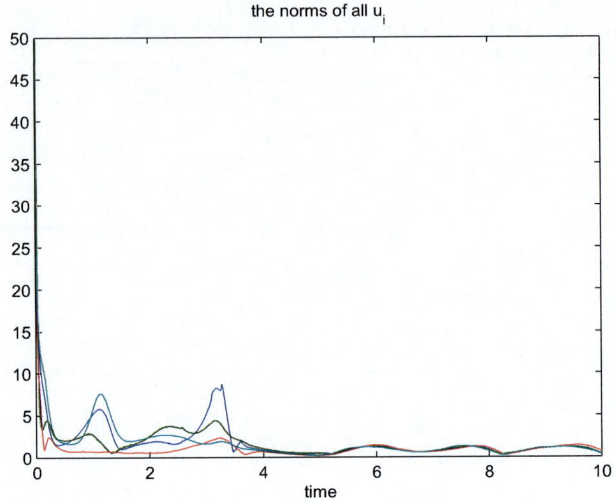

Fig. 7.4 Time responses of the norms of four followers' input control signals (7.21)

7.3.2 *Adaptive Sliding-mode Attitude Containment Control*

The simulation result under the adaptive sliding-mode attitude coordination controller (7.29)–(7.30) will be shown in this section. The initial values of all rigid bodies' orientations and angular velocities, external disturbances and actuation effectiveness matrices are the same as those in Sect. 7.3.1. The parameters in (7.29)–(7.30) are chosen as

$$\begin{aligned}
&k_{12} = 2, k_{22} = 7, k_{32} = 3, k_{42} = 2\\
&k_{13} = 1, k_{23} = 5, k_{33} = 2, k_{43} = 5\\
&\rho_1 = 0.6, \rho_2 = 0.7, \rho_3 = 0.9, \rho_4 = 0.8\\
&\widetilde{\alpha}_1 = 0.9, \widetilde{\alpha}_2 = 0.9, \widetilde{\alpha}_3 = 0.9, \widetilde{\alpha}_4 = 0.8\\
&a_1 = 4, a_2 = 3, a_3 = 3, a_4 = 2\\
&\varepsilon_1 = 0.5, \varepsilon_2 = 1, \varepsilon_3 = 1, \varepsilon_4 = 1\\
&k_1 = k_2 = k_3 = k_4 = 1
\end{aligned}$$

and the parameters $\beta_i (i = 1, 2, 3, 4)$ are assumed to be

$$\beta_i = \left(\frac{k_{i2}}{2a_i\widetilde{\alpha}_i}\right)^{\frac{2}{1-\rho_i}} \left(\frac{2}{\lambda_{i,max}}\right)^{\frac{1+\rho_i}{1-\rho_i}}$$

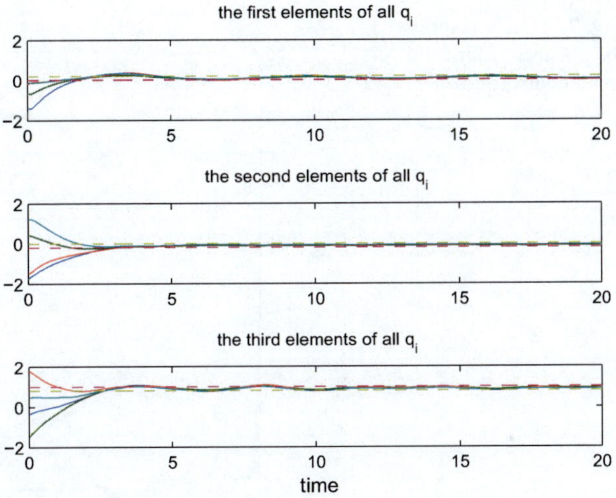

Fig. 7.5 Time responses of all rigid bodies' orientations under controllers (7.29)–(7.30)

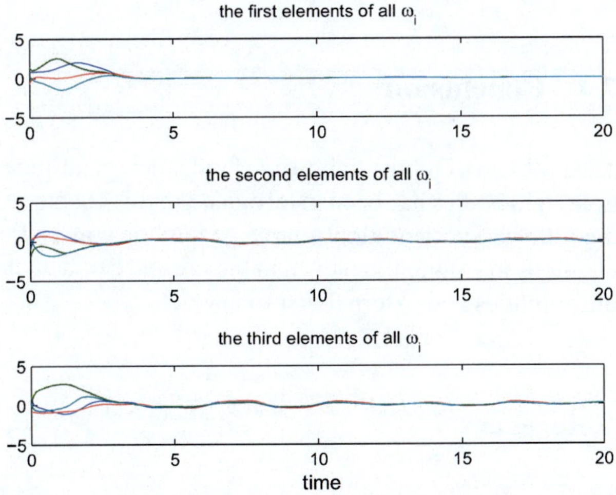

Fig. 7.6 Time responses of all rigid bodies' angular velocities under controllers (7.29)–(7.30)

The simulation results are shown in Figs. 7.5, 7.7 and 7.7. The dash lines and solid lines denote the time responses of the leaders' states and the followers' states respectively in Figs. 7.5 and 7.6. It can be seen that the control task is accomplished in a finite time. The norms of the input control signals of all four followers are depicted in Fig. 7.7.

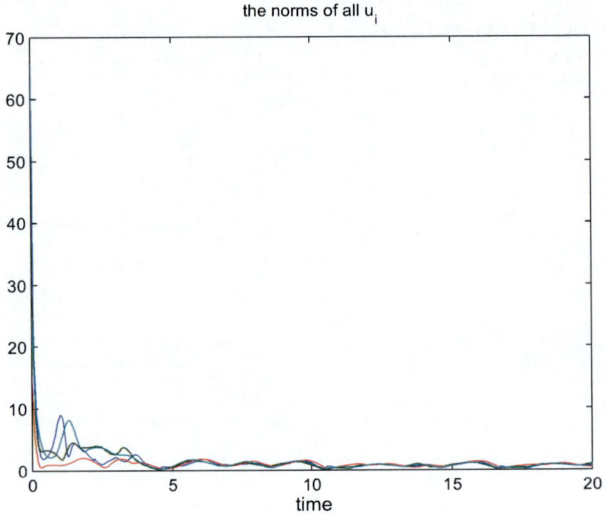

Fig. 7.7 Time responses of the norms of four followers' input control signals (7.29)

7.4 Conclusion

The problem of robust finite-time fault-tolerant attitude containment control for multiple rigid bodies has been investigated in this chapter. Under the designed controller, the followers' orientations converge into the convex hull formed by the leaders' orientations in a finite time, even in the presence of actuator failures, system parametric uncertainties and external disturbances.

References

1. J.L. Junkins, H. Schaub, *Analytical Mechanics of Space Systems* (American Institute of Aeronautics and Astronautics, 2009)
2. Y. Hong, J. Hu, L. Gao, Tracking control for multi-agent consensus with an active leader and variable topology. Automatica **42**(7), 1177–1182 (2006)
3. Z. Meng, W. Ren, Z. You, Distributed finite-time attitude containment control for multiple rigid bodies. Automatica **46**(12), 2092–2099 (2010)
4. S.P. Bhat, D.S. Bernstein, Finite-time stability of continuous autonomous systems. SIAM J. Control Optim. **38**(3), 751–766 (2000)
5. Z. Zhu, Y. Xia, M. Fu, Attitude stabilization of rigid spacecraft with finite-time convergence. Int. J. Robust Nonlinear Control **21**(6), 686–702 (2011)
6. P.G. Hardy, J Littlewood, *Inequalities* (Cambridge, U.K., Cambridge University Press, 1952)
7. A.-M. Zou, K.D. Kumar, Robust attitude coordination control for spacecraft formation flying under actuator failures. J. Guidance Control Dyn. **35**(4), 1247–1255 (2012)
8. S. Yu, X. Yu, B. Shirinzadeh, Z. Man, Continuous finite-time control for robotic manipulators with terminal sliding mode. Automatica **41**(11), 1957–1964 (2005)
9. D.V. Dimarogonas, P. Tsiotras, K.J. Kyriakopoulos, Leader-follower cooperative attitude control of multiple rigid bodies. Syst. Control Lett. **58**(6), 429–435 (2009)

Part IV
Distributed Data-Driven Optimal Consensus Control for Multi-agent Systems

Chapter 8
Data-Driven Optimal Consensus Control for Linear Multiagent Systems

This chapter is concerned with data-driven distributed optimal consensus control for unknown MASs with input delays. The input-delayed MAS model is firstly converted into a delay-free form using a model reduction method. By establishing an equivalent relationship on the predesigned performance indices of the two MASs, optimal consensus control of input-delayed MAS can be fully transformed to that of delay-free MAS. Based on the coupled Hamilton-Jacobi equations and Bellman's optimality principle, optimal consensus control policies are derived for the transformed delay-free MAS. Then a policy iteration algorithm based on distributed asynchronous update mechanism is proposed to learn the coupled Hamilton-Jacobi-Bellman equations online. To perform the proposed data-driven adaptive dynamic programming algorithm, we adopt the measured data based critic-actor neural networks to approximate the value functions and the control policies, respectively. Finally, a simulation example is given to illustrate the effectiveness of the proposed method.

The rest of this paper is organized as follows. Section 8.1 presents problem formulation and transformation which includes system model transformation and the equivalence of the performance indexes. Distributed optimal consensus controllers are designed in Sect. 8.2 while the PI algorithm is introduced to online learn the HJB equations which inspires our work in Sect. 8.3. We propose critic-actor NNs structure to approximate the value functions and control policies online based on the gradient descending method in Sect. 8.4. A simulation example is given to illustrate the effectiveness of the proposed method in Sect. 8.5. Finally, we conclude our scheme briefly.

D. Yue et al., *Distributed Cooperative Control and Communication for Multi-agent Systems*, https://doi.org/10.1007/978-981-33-6718-0_8

8.1 Problem Formulation

Consider an input-delayed MAS with N agents, each with dynamics described as

$$\dot{x}_i(t) = Ax_i(t) + B_i u_i(t - \tau_i) \tag{8.1}$$

where $x_i(t) \in R^n$ and $u_i(t) \in R^m$ denote the state and control input for agent i, respectively. A and B_i are unknown constant matrices with appropriate dimensions. τ_i is a bounded input delay and satisfies $\tau_i \in [0, h_i T_s]$ with T_s being sampling periodic. $x_i(\theta), \theta \in [-\tau_i, 0]$ is given and bounded.

Assumption 8.1 Suppose that the sensors, controllers, actuators are time-driven and the transmitted packets are ordered.

In order to investigate consensus control of system (8.1), we define the local neighbor consensus error for agent i as

$$\delta_i(t) = \sum_{j \in N_i} a_{ij}(x_i(t) - x_j(t)) \tag{8.2}$$

Taking the derivative of (8.2) with respect to time t, the local neighbor consensus error dynamics for agent i can be represented as

$$\begin{aligned} \dot{\delta}_i(t) &= \sum_{j \in N_i} a_{ij}(\dot{x}_i(t) - \dot{x}_j(t)) \\ &= A\delta_i(t) + d_i B_i u_i(t) - \sum_{j \in N_i} a_{ij} B_j u_j(t) \end{aligned} \tag{8.3}$$

In the distributed control scheme, agent i depends on the information of itself and its neighbors. Then for MAS (8.1), the performance index for agent i is affected by the control inputs of itself and its neighbors as well as the local neighbor consensus error, so we define a local performance index function of agent i as

$$\begin{aligned} &J_x^i(\delta_i(0), U_{\bar{N}_i}) \\ &= \int_0^\infty r_x^i(\delta_i(t), U_{\bar{N}_i}(t))dt \\ &= \int_0^\infty \left[U_i^T(t)\ U_1^T(t) \ldots U_{N_i}^T(t)\ \delta_i^T(t)\right] T^i \left[U_i^T(t)\ U_1^T(t) \ldots U_{N_i}^T(t)\ \delta_i^T(t)\right]^T dt \end{aligned} \tag{8.4}$$

where $r_x^i(\delta_i(t), U_{\bar{N}_i}(t))$ is the utility function with respect to system (8.1), $U_{\bar{N}_i}(t) = \{U_j(t) : U_j(t) = [u_j((k - h_j)T_s)^T, \ldots, u_j(kT_s)^T]^T, j \in \bar{N}_i\}$ denotes the whole control input vector in $[kT_s, (k+1)T_s]$, and the matrix T^i is symmetric positive semi-definite matrix. The cost functional (8.4) can be regarded as a local performance index on energy consumption during consensus evolution of MASs.

From the above analysis, we propose the following problem:

The problem required to be solved is that how to design distributed optimal consensus control protocols which not only make all the agents be able to reach consensus but minimize the performance indices (8.4) as well.

8.2 Problem Transformation

In order to simplify the discussion of optimal consensus problem for system (8.1), a model reduction method is introduced to deal with input delays and then the model of input-delayed MAS can be converted into a delay-free MAS form.

Introducing a new state variable

$$z_i(t) = x_i(t) + \int_t^{t+\tau_i} e^{A(t-\eta)} B_i u_i(\eta - \tau_i) d\eta \tag{8.5}$$

and differentiating it with respect to time t, we can reduce (8.1) to a delay-free MAS form

$$\begin{aligned} \dot{z}_i(t) &= \dot{x}_i(t) + A \int_t^{t+\tau_i} e^{A(t-\eta)} B_i u_i(\eta - \tau_i) d\eta + e^{-A\tau_i} B_i u_i(t) - B_i u_i(t - \tau_i) \\ &= Ax_i(t) + A \int_t^{t+\tau_i} e^{A(t-\eta)} B_i u_i(\eta - \tau_i) d\eta + e^{-A\tau_i} B_i u_i(t) \\ &= Az_i(t) + C_i u_i(t) \end{aligned} \tag{8.6}$$

where $C_i \triangleq e^{-A\tau_i} B_i$.

Similar to (8.2), we define the local neighbor consensus error for the transformed delay-free system (8.6) as

$$e_i(t) = \sum_{j \in N_i} a_{ij}(z_i(t) - z_j(t)) \tag{8.7}$$

And differentiating (8.7) with respect to time t, we yield

$$\begin{aligned} \dot{e}_i(t) &= \sum_{j \in N_i} a_{ij}(\dot{z}_i(t) - \dot{z}_j(t)) \\ &= Ae_i(t) + d_i C_i u_i(t) - \sum_{j \in N_i} a_{ij} C_j u_j(t) \end{aligned} \tag{8.8}$$

According to [1, 2], we select a local performance index function of system (8.6) as

$$\begin{aligned} J_z^i(e_i(0), u_{\bar{N}_i}) &= \int_0^\infty r_z^i(e_i(t), u_{\bar{N}_i}(t))dt \\ &= \int_0^\infty (e_i^T(t)S_{ii}e_i(t) + u_i^T(t)L_{ii}u_i(t) + \sum_{j\in N_i} u_j^T(t)L_{ij}u_j(t))dt \end{aligned} \tag{8.9}$$

for agent i where $r_z^i(e_i(t), u_{\bar{N}_i}(t))$ is the utility function with respect to (8.6), $u_{\bar{N}_i} = \{u_j : j \in \bar{N}_i\}$ denotes the control input for agent i and its neighbors, and S_{ii}, L_{ii}, L_{ij} are symmetric positive definite matrices.

In order to build an equivalent relationship on optimal consensus problem between MAS (8.4) and MAS (8.9), we introduce the following lemma.

Lemma 8.1 *Let Assumption 8.1 hold, then the new state variable z_i can be represented as*

$$\begin{aligned} z_i(t) &= x_i(t) + e^{A(t-\tau_i)}(F^i_{k+1-m_i,h_i}u_i((k-h_i)T_s) + \cdots \\ &+ F^i_{k+1-m_i,q_i}u_i((k-q_i)T_s) + \cdots + F^i_{k+1-m_i,0}u_i(kT_s)) \\ &= x_i(t) + F^iU_i(t) \end{aligned}$$

where $t-\tau_i \in [(k-m_i)T_s, (k+1-m_i)T_s]$ with T_s being the sampling period, and

$$F^i = e^{A(t-\tau_i)}[F^i_{k+1-m_i,h_i}, \ldots, F^i_{k+1-m_i,q_i}, \ldots, F^i_{k+1-m_i,0}]$$

$$F^i_{k+1,0} = \int_{t-\tau_i}^{t} e^{-Av}dvB_i$$

$$F^i_{k+1-m_i,q_i} = \begin{cases} \int_{kT_s}^{t} e^{-Av}dvB_iu_i(kT_s), q_i = 0; \\ \int_{(k-q_i)T_s}^{(k+1-q_i)T_s} e^{-Av}dvB_iu_i((k-q_i)T_s), q_i = 1, \ldots, m_i - 1; \\ \int_{t-\tau_i}^{(k+1-m_i)T_s} e^{-Av}dvB_iu_i((k-m_i)T_s), q_i = m_i; \\ 0, q_i = m_i + 1, \ldots, h_i. \end{cases}$$

with $m_i = 0, \ldots, h_i$.

Proof Define $\nu = \eta - \tau_i$, then we can obtain

$$\int_{t}^{t+\tau_i} e^{A(t-\eta)} B_i u_i(\eta - \tau_i) d\eta$$

$$= \int_{t-\tau_i}^{t} e^{A(t-\tau_i-\nu)} B_i u_i(\nu) d\nu$$

$$= e^{A(t-\tau_i)} \int_{t-\tau_i}^{t} e^{-A\nu} B_i u_i(\nu) d\nu$$

According to Assumption 8.1, there exists a constant $k \geq 0$ such that $t \in [kT_s, (k+1)T_s]$. Assume that the input delay satisfies $\tau_i \in [0, h_i T_s]$, we discuss $\int_{t-\tau_i}^{t} e^{-A\nu} B_i u_i(\nu) d\nu$ as follows:

(1) When $\tau_i \in [0, T_s]$, we have $t - \tau_i \in [(k-1)T_s, (k+1)T_s]$. Thus we can obtain

$$\int_{t-\tau_i}^{t} e^{-A\nu} B_i u_i(\nu) d\nu$$

$$= \begin{cases} \int_{t-\tau_i}^{kT_s} e^{-A\nu} d\nu B_i u_i((k-1)T_s) + \int_{kT_s}^{t} e^{-A\nu} d\nu \\ \times B_i u_i(kT_s), \qquad t - \tau_i \in [(k-1)T_s, kT_s]; \\ \int_{t-\tau_i}^{t} e^{-A\nu} d\nu B_i u_i(kT_s), \quad t - \tau_i \in [kT_s, (k+1)T_s]. \end{cases}$$

$$\triangleq \begin{cases} F_{k,1}^i u_i((k-1)T_s) \\ +F_{k,0}^i u_i(kT_s), \qquad t - \tau_i \in [(k-1)T_s, kT_s]; \\ F_{k+1,0}^i u_i(kT_s), \qquad t - \tau_i \in [kT_s, (k+1)T_s]. \end{cases}$$

(2) when $\tau_i \in [(l-1)T_s, lT_s]$, then one has $t - \tau_i \in [(k-l)T_s, (k+2-l)T_s]$ where $l = 2, \ldots, h_i$. Thus we can obtain

$$\int_{t-\tau_i}^{t} e^{-A\nu} B_i u_i(\nu) d\nu$$

$$
= \begin{cases} \int_{t-\tau_i}^{(k+1-l)T_s} e^{-Av} dv B_i u_i((k-l)T_s) + \cdots \\ + \int_{(k+p-l)T_s}^{k+p+1-l)T_s} e^{-Av} dv B_i u_i((k+p-l)T_s) + \cdots \\ + \int_{kT_s}^{t} e^{-Av} dv B_i u_i(kT_s), \quad p = 1, \ldots, l \; and \\ \quad t - \tau_i \in [(k-l)T_s, (k+1-l)T_s]; \\ \int_{t-\tau_i}^{(k+2-l)T_s} e^{-Av} dv B_i u_i((k+1-l)T_s) + \cdots \\ + \int_{(k+p-l)T_s}^{k+p+1-l)T_s} e^{-Av} dv B_i u_i((k+p-l)T_s) + \cdots \\ + \int_{kT_s}^{t} e^{-Av} dv B_i u_i(kT_s), \quad p = 2, \ldots, l \; and \\ \quad t - \tau_i \in [(k+1-l)T_s, (k+2-l)T_s]. \end{cases}
$$

$$
\triangleq \begin{cases} F^i_{k+1-l,l} u_i((k-l)T_s) + \cdots + F^i_{k+1-l,l-p} \\ \times u_i((k+p-l)T_s) + \cdots + F^i_{k+1-l,0} u_i(kT_s), \; p = 1, \\ \ldots, l \; and \; t - \tau_i \in [(k-l)T_s, (k+1-l)T_s]; \\ F^i_{k+2-l,l-1} u_i((k+1-l)T_s) + \cdots + F^i_{k+2-l,l-p} \\ \times u_i((k+p-l)T_s) + \cdots + F^i_{k+2-l,0} u_i(kT_s), \; p = 2, \\ \ldots, l \; and \; t - \tau_i \in [(k+1-l)T_s, (k+2-l)T_s]. \end{cases}
$$

Based on the above analysis, when the time intervals satisfy $t - \tau_i \in [(k - m_i)T_s, (k+1-m_i)T_s]$ with $m_i = 0, \ldots, h_i$, then we have

$$
\begin{aligned} \int_{t-\tau_i}^{t} e^{-Av} B_i u_i(v) dv &= F^i_{k+1-m_i,m_i} u_i((k-m_i)T_s) + \cdots + F^i_{k+1-m_i,0} u_i(kT_s) \\ &= F^i_{k+1-m_i,h_i} u_i((k-h_i)T_s) + \cdots + F^i_{k+1-m_i,m_i} \\ &\quad \times u_i((k-m_i)T_s) + \cdots + F^i_{k+1-m_i,0} u_i(kT_s) \end{aligned}
$$

Thus according to the definition of the new state variable $z_i(t)$, we can obtain

$$
\begin{aligned} z_i(t) &= x_i(t) + e^{A(t-\tau_i)} (F^i_{k+1-m_i,h_i} u_i((k-h_i)T_s) + \cdots \\ &\quad + F^i_{k+1-m_i,q_i} u_i((k-q_i)T_s) + \cdots + F^i_{k+1-m_i,0} u_i(kT_s)) \\ &= x_i(t) + F^i U_i(t) \end{aligned}
$$

where $F^i = e^{A(t-\tau_i)}[F^i_{k+1-m_i,h_i}, \ldots, F^i_{k+1-m_i,q_i}, \ldots, F^i_{k+1-m_i,0}]$, $F^i_{k+1,0} = \int_{t-\tau_i}^{t} e^{-Av} dv B_i$ and

$$F^i_{k+1-m_i,q_i} = \begin{cases} \int\limits_{kT_s}^{t} e^{-Av} dv B_i u_i(kT_s), q_i = 0; \\ \int\limits_{(k-q_i)T_s}^{(k+1-q_i)T_s} e^{-Av} dv B_i u_i((k-q_i)T_s), q_i = 1, \ldots, m_i - 1; \\ \int\limits_{t-\tau_i}^{(k+1-m_i)T_s} e^{-Av} dv B_i u_i((k-m_i)T_s), q_i = m_i; \\ 0, q_i = m_i + 1, \ldots, h_i. \end{cases}$$

Theorem 8.1 *Consider a directed connected MAS (8.1) which contains a directed spanning tree and let Assumption 8.1 hold. Assume that there exists a global state feedback controller* $U(t) = Ke(t)$ *with* $U(t) = [u_1^T(t), \ldots, u_N^T(t)]^T$ *and* $e(t) = [e_1^T(t), \ldots, e_N^T(t)]^T$ *which can make the transmitted delay-free MAS (8.6) achieve consensus, then the original input-delayed MAS (8.1) can also reach consensus with the same controller.*

Proof From (8.2), (8.7) and (8.10), we have

$$\begin{aligned} e_i(t) &= \sum_{j\in N_i} a_{ij}(z_i(t) - z_j(t)) \\ &= \sum_{j\in N_i} a_{ij}(x_i(t) + F^i U_i(t) - x_j(t) - F^j U_j(t)) \\ &= \delta_i(t) + d_i F^i U_i(t) - \sum_{j\in N_i} a_{ij} F^j U_j(t) \\ &= \delta_i(t) + \sum_{j\in \bar{N}_i} h_{ij} F^j U_j(t) \end{aligned} \tag{8.10}$$

Let the global error vector of the original input-delayed MAS be $\delta(t)=[\delta_1^T(t), \ldots, \delta_N^T(t)]^T$, then the compact form of (8.10) can be written as

$$e(t) = \delta(t) + (H \otimes I_n)\mathcal{F}\mathbb{U}(t) \tag{8.11}$$

where $\mathbb{U}(t) = [U_1^T, U_2^T, \ldots, U_N^T]^T$ and $\mathcal{F} = [F^1, \ldots, F^N]$.

With the definitions of $\mathbb{U}$ and U, it easily yields $\|\mathbb{U}\| \leq \bar{h}\|U\|$ where $\bar{h} = \max\{h_i | i = 1, \ldots, N\}$.

Thus we can obtain

$$\begin{aligned} \|\delta(t)\| &= \|e(t) - (H \otimes I_n)\mathcal{F}\mathbb{U}(t)\| \\ &\leq \|e(t)\| + \|(H \otimes I_n)\mathcal{F}\mathbb{U}(t)\| \\ &\leq \|e(t)\| + \bar{h}\|(H \otimes I_n)\mathcal{F}K\|\|e(t)\| \\ &= (1 + \bar{h}\|(H \otimes I_n)\mathcal{F}K\|)\|e(t)\| \end{aligned} \tag{8.12}$$

then $\delta(t) \to 0$ as $e(t) \to 0$. In other words, system (8.1) can also reach consensus while system (8.6) achieves consensus.

Theorem 8.2 *Consider the performance indices (8.4) and (8.9) of the two MASs (8.1) and (8.6). Under Assumption 8.1, if there exist matrices $T^i \geq 0$, $S_{ii} > 0$, $L_{ii} > 0$, and $L_{ij} > 0$, $j \in N_i$ satisfying*

$$T^i = \begin{bmatrix} I_1^T L_{i1} I_1 & & & & & \\ & \ddots & & & & \\ & & I_1^T L_{is} I_1 & & & \\ & & & 0 & & \\ & & & & \ddots & \\ & & & & & 0 \\ & & & & & & S_{ii} \end{bmatrix}$$

where $I_1 = [1, 0, \ldots, 0]$, then the performance indices for the input-delayed MAS (8.1) are equivalent to those of the delay-free MAS (8.6), i.e., $J_x^i = J_z^i$, $\forall i$.

Proof From (8.11), we obtain

$$\begin{aligned} e_i^T(t) S_{ii} e_i(t) &= (\delta_i(t) + \sum_{j\in\bar{N}_i} h_{ij} F^j U_j(t))^T S_{ii} (\delta_i(t) + \sum_{j\in\bar{N}_i} h_{ij} F^j U_j(t)) \\ &= \delta_i^T(t) S_{ii} \delta_i(t) + \sum_{j\in\bar{N}_i} h_{ij} \delta_i^T(t) S_{ii} F^j U_j(t) + \sum_{j\in\bar{N}_i} h_{ij} U_j^T(t) \\ &\quad \times F^{jT} S_{ii} \delta_i(t) + \sum_{j\in\bar{N}_i} h_{ij}^2 U_j^T(t) F^{jT} S_{ii} F^j U_j(t) \\ &\quad + \sum_{p=1}^{N-1} \sum_{k=p+1}^{N} h_{ip} h_{ik} U_p^T(t) F^{pT} S_{ii} F^k U_k(t) \\ &\quad + \sum_{k=2}^{N} \sum_{p=1}^{k-1} h_{ik} h_{ip} U_k^T(t) F^{kT} S_{ii} F^p U_p(t) \end{aligned}$$

where the parameters $p, k \in \bar{N}_i$.

Because of $u_i(t) = [1, 0, \ldots, 0] U_i(t) = I_1 U_i(t)$, then one has $u_i^T(t) L_{ii} u_i(t) = U_i^T(t) I_1^T L_{ii} I_1 U_i(t)$ and $u_j^T(t) L_{ij} u_j(t) = U_j^T(t) I_1^T L_{ij} I_1 U_j(t)$.

Thus the utility function of (8.9) is rewritten as

$$\begin{aligned} & r_z^i(e_i(t), u_{\bar{N}_i}(t)) \\ &= (\delta_i(t) + \sum_{j\in\bar{N}_i} h_{ij} F^j U_j(t))^T S_{ii} (\delta_i(t) + \sum_{j\in\bar{N}_i} h_{ij} F^j U_j(t)) \end{aligned}$$

$$= \delta_i^T(t)S_{ii}\delta_i(t) + \sum_{j\in\bar{N}_i} h_{ij}\delta_i^T(t)S_{ii}F^jU_j(t) + \sum_{j\in\bar{N}_i} h_{ij}U_j^T(t)$$
$$\times F^{jT}S_{ii}\delta_i(t) + \sum_{j\in\bar{N}_i} \bar{a}_{ij}^2 U_j^T(t)F^{jT}S_{ii}F^jU_j(t)$$
$$+ \sum_{p=1}^{N-1}\sum_{k=p+1}^{N} h_{ip}h_{ik}U_p^T(t)F^{pT}S_{ii}F^kU_k(t)$$
$$+ \sum_{k=2}^{N}\sum_{p=1}^{k-1} h_{ik}h_{ip}U_k^T(t)F^{kT}S_{ii}F^pU_p(t)$$
$$+ U_i^T(t)I_1^TL_{ii}I_1U_i(t) + \sum_{j\in N_i} U_j^T(t)I_1^TL_{ij}I_1U_j(t)$$

And from (8.4), we also have

$$r_x^i(e_i(t), U_{\bar{N}_i}(t)) = \sum_{j\in\bar{N}_i} U_j^T(t)T_{j,j}^iU_j(t) + \sum_{p=1}^{N-1}\sum_{k=p+1}^{N} U_p^T(t)T_{p,k}^iU_k(t)$$
$$+ \sum_{k=2}^{N}\sum_{p=1}^{k-1} U_k^T(t)T_{k,p}^iU_p(t) + \delta_i^T(t)T_{N+1,N+1}^i\delta_i(t)$$
$$+ \sum_{j\in\bar{N}_i} \delta_i^T(t)T_{N+1,j}^iU_j(t) + \sum_{j\in\bar{N}_i} U_j^T(t)T_{j,N+1}^i\delta_i(t)$$

In order to satisfy $r_x^i(t) = r_z^i(t)$, the following inequalities must hold:

$$\begin{cases} T_{jj}^i = 0,\ j \notin N_i \\ T_{jj}^i = h_{ij}^2F^{jT}S_{ii}F^j + I_1^TL_{ij}I_1,\ j \in N_i \\ T_{p,k}^i = h_{ip}h_{ik}F^{pT}S_{ii}F^k,\ p = 1,\dots,N-1;\ k = p+1,\dots,N \\ T_{k,p}^i = h_{ik}h_{ip}F^{kT}S_{ii}F^p,\ k = 2,\dots N;\ p = 1,\dots,k-1 \\ T_{N+1,j}^i = h_{ij}S_{ii}F^j,\ j = 1,\dots,N \\ T_{j,N+1}^i = h_{ij}F^{jT}S_{ii},\ j = 1,\dots,N \\ T_{N+1,N+1}^i = S_{ii} \end{cases} \tag{8.13}$$

Through linear transformation, we obtain

$$T^i = \begin{bmatrix} I_1^T L_{i1} I_1 & & & & & \\ & \ddots & & & & \\ & & I_1^T L_{is} I_1 & & & \\ & & & 0 & & \\ & & & & \ddots & \\ & & & & & 0 \\ & & & & & & S_{ii} \end{bmatrix} \geq 0$$

Thus when the conditions (8.13) are satisfied, the performance indexes for the two MASs are equivalent, i.e., $J_x^i = J_z^i, \forall i$.

Based on Theorem 8.1, the linear controller $U(t) = Ke(t)$ stabilizes the transformed local neighborhood consensus error system (8.8), then the original local neighborhood consensus error system (8.3) is also stable with the same controller. That is to say, the input-delayed MAS (8.1) achieves synchronization while the delay-free MAS (8.6) reaches consensus. Moreover, Theorem 2 shows that the performance indices of the two MAS (8.1) and (8.6) are equivalent. Thus it is easily seen that the optimal consensus control problem of system (8.1) can be completely converted to that of system (8.6). In the following, optimal consensus control policies will be designed to make all the agents with system (8.6) reach consensus and to minimize the performance indices (8.9).

8.3 Optimal Consensus Control Design

For the delay-free MAS (8.6), optimal consensus control problem is similar to multi-player cooperative games [1, 2]. Thus the machinery of N-player cooperative games is utilized to design optimal consensus controller of system (8.6) in the section.

8.3.1 *Optimal Consensus Control for Delay-free Multiagent Systems*

Definition 8.1 (*Admissible consensus control policy* [1, 3]) Feedback control policy $u_i, \forall i$ is said to be admissible consensus control with respect to (8.9) on a set $\Pi \in \mathbb{R}^n$ if u_i is continuous, $u_i(0) = 0$, u_i stabilizes the system (8.8) locally, and the local cost functional (8.9) is finite.

Given the admissible consensus control policies u_j, $j \in \bar{N}_i$, the local value function $V_i(e_i(t))$ for agent i is defined as

$$V_i(e_i(t)) = \int_t^\infty r_z^i(e_i(\tau), u_{\bar{N}_i}(\tau))d\tau \tag{8.14}$$
$$= \int_t^\infty \left(e_i^T S_{ii} e_i + u_i^T L_{ii} u_i + \sum_{j \in N_i} u_j^T L_{ij} u_j \right) d\tau$$

and the associated local coupled HJ equation is represented as

$$\begin{aligned} H_i(e_i, u_{\bar{N}_i}, V_{e_i}) &= r_z^i(e_i, u_{\bar{N}_i}) + V_{e_i}^T \dot{e}_i \\ &= e_i^T S_{ii} e_i + u_i^T L_{ii} u_i + \sum_{j \in N_i} u_j^T L_{ij} u_j \\ &\quad + V_{e_i}^T \left(A e_i + d_i C_i u_i - \sum_{j \in N_i} a_{ij} C_j u_j \right) \end{aligned} \tag{8.15}$$

where $V_{e_i}(t) = \partial V_i(e_i(t))/\partial e_i(t)$ is the derivative of the value functions $V_i(e_i(t))$ with respect to $e_i(t)$.

Solving the first order necessary condition $\partial H_i(t)/\partial u_i(t) = 0$, we have

$$u_i(t) = -\frac{1}{2} d_i L_{ii}^{-1} C_i^T V_{e_i} \tag{8.16}$$

According to Bellman's optimality principle, the optimal local value function $V_i^*(e_i(t))$ satisfies the local coupled HJB equation

$$\min H_i(e_i, u_i^*, u_{N_i}, V_{e_i}^*) = 0 \tag{8.17}$$

Thus optimal consensus control policy for agent i is

$$u_i^*(t) = -\frac{1}{2} d_i L_{ii}^{-1} C_i^T V_{e_i}^* \tag{8.18}$$

According to [1], there exists a positive definition matrix $P \in R^{nN \times nN}$ such that

$$V_e(t) = Pe(t) \tag{8.19}$$

where $V_e(t) = [V_{e_1}^T(t), \ldots, V_{e_N}^T(t)]^T$.

Based on (8.16), the overall control input can be represented as

$$U(t) = -\frac{1}{2} D \mathfrak{L}^{-1} \mathfrak{C}^T P e(t) \triangleq K e(t) \tag{8.20}$$

where $K = -\frac{1}{2} D \mathfrak{L}^{-1} \mathfrak{C}^T P$, $\mathfrak{L} = diag\{L_{ii}\}$ and $\mathfrak{C} = diag\{C_i\}$. Thus the designed global optimal consensus controller U is linear with respect to $e(t)$.

Remark 8.1 In the above analysis, the overall control input $U(t)$ depends on the global neighbor consensus error $e(t)$. From (8.12), meanwhile, it is easily known that $e(t)$ is a combination of $\delta(t)$ and the past and current control inputs $\mathbb{U}(t)$. Thus the global control input $U(t)$ relies on the neighbor consensus error $\delta(t)$ and the control inputs $\mathbb{U}(t)$.

The linear controller (8.20) can make the transformed delay-free MAS (8.6) achieve optimal consensus. According to Theorem 8.1 and 8.2, it is obtained that the original input-delayed MAS (8.1) reach optimal consensus as well.

8.3.2 Nash Equilibrium

Definition 8.2 (*Global Nash Equilibrium* [2]) An N-tuple of control policies $\{u_1^*, u_2^*, \ldots, u_i^*, \ldots, u_N^*\}$ is said to be a global Nash equilibrium solution for an N-player game if there exists

$$\begin{aligned} J_i^* &\triangleq J_i(u_1^*, u_2^*, \ldots, u_i^*, \ldots, u_N^*) \\ &\leq J_i(u_1^*, u_2^*, \ldots, u_i, \ldots, u_N^*) \end{aligned} \tag{8.21}$$

where J_i denotes the performance of agent i, $i = 1, \ldots, N$.

The N-tuple of the local performance indices $\{J_1^*, J_2^*, \ldots, J_N^*\}$ is called as a Nash equilibrium of the N-player game.

Theorem 8.3 *Consider the local neighbor consensus error system (8.8), when the value functions $V_i^*(e_i(t))$ are the solution to the coupled HJB equations (8.17), and optimal consensus control policies u_i^* are chosen as (8.18), then system (8.8) is asymptotically stable. Meanwhile, $J_z^{i*}(e_i(0), u_i^*, u_{N_i}^*) = V_i^*(e_i(0))$, u_i^* and $u_{N_i}^*$ are in Nash equilibrium.*

Proof Consider $V_i^*(e_i)$ as the Lyapunov function candidate, then we can easily obtain

$$\begin{aligned} \dot{V}_i^*(e_i) &= V_{e_i}^{*T} \dot{e}_i \\ &= V_{e_i}^{*T}(Ae_i(t) + d_i C_i u_i^*(t) - \sum_{j \in N_i} a_{ij} C_j u_j^*(t)) \\ &= -e_i^T S_{ii} e_i - u_i^{*T} L_{ii} u_i^* - \sum_{j \in N_i} u_j^{*T} L_{ij} u_j^* < 0 \end{aligned} \tag{8.22}$$

Thus system (8.8) is asymptotically stable.

According to (8.9) and (8.14), it is apparent that $J_z^{i*}(e_i(0), u_i^*, u_{N_i}^*) = V_i^*(e_i(0))$, u_i^* and $u_{N_i}^*$ are in Nash equilibrium.

8.4 Policy Iteration Algorithm

In this section, a policy iteration (PI) algorithm is introduced to learn the coupled HJB equations online. In order to reduce the computational burden in comparison with the synchronous scheme in [2, 4, 5], we propose a distributed asynchronous mechanism to update the control policies in the PI algorithm. To study the update of agent i, the neighbor agents N_i are classified into 2 teams and then let us consider the update case of them. Assume that $N_i = N_i^1 + N_i^2$ in which N_i^1 denotes the synchronous updated neighbor agents while N_i^2 is the other neighbors.

PI Algorithm:
Step 1: Initialization. Start with arbitrary initial admissible control policies $u_i^{[0]}, \forall i$.
Step 2: Policy Evaluation. Given the iterative control policies $u_i^{[k]}, u_{N_i^1}^{[k]}, u_{N_i^2}^{[k-1]}$, solve for N-tuple of costs $V_i^{[k]}, \forall i$ using HJ equations

$$0 = H_i(e_i, V_{e_i}^{[k]}, u_i^{[k]}, u_{N_i^1}^{[k]}, u_{N_i^2}^{[k-1]}) \tag{8.23}$$

Step 3: Policy Improvement. Update the N-tuple of iterative control policies using

$$u_i^{[k+1]} = -\frac{1}{2} d_i L_{ii}^{-1} C_i^T V_{e_i}^{[k]} \tag{8.24}$$

Step 4: On convergence,end. Else $k = k + 1$ and go back to Step 2.

Theorem 8.4 *Given fixed neighbor control policies $u_{N_i^2}$ and assume there exist admissible control policies $u_{N_i^1}$. Suppose that the agents i and N_i^1 perform PI algorithm synchronously while the agents N_i^2 do not update their control policies. Then for small $\lambda_{min}(L_{ij})$ and big $\lambda_{max}(L_{jj}^{-T} L_{ij})$, $u_i, \forall i$ converges to the Nash Equilibrium, and the value function $V_i^k, \forall i$ converges to optimal value function V_i^*.*

Proof For agent i, we have

$$\begin{aligned}
&H_i(e_i, V_{e_i}^{[k+1]}, u_i^{[k+1]}, u_{N_i^1}^{[k+1]}, u_{N_i^2}^{[k]}) - H_i(e_i, V_{e_i}^{[k]}, u_i^{[k]}, u_{N_i}^{[k]}) \\
&= \sum_{j \in \{N_i^1, i\}} (u_j^{[k+1]T} L_{ij} u_j^{[k+1]} - u_j^{[k]T} L_{ij} u_j^{[k]}) + \Xi_i \\
&= \sum_{j \in \{N_i^1, i\}} (u_j^{[k+1]} - u_j^{[k]})^T L_{ij} (u_j^{[k+1]} - u_j^{[k]}) \\
&\quad + 2 \sum_{j \in \{N_i^1, i\}} u_j^{[k]T} L_{ij} (u_j^{[k+1]} - u_j^{[k]}) + \Xi_i \\
&= r_z^i(e_i, u_i^{[k+1]}, u_{N_i^1}^{[k+1]}, u_{N_i^2}^{[k]}) - r_z^i(e_i, u_i^{[k]}, u_{N_i}^{[k]}) + \Xi_i
\end{aligned} \tag{8.25}$$

where $\Xi_i = \dot{V}_i^{[k+1]} - \dot{V}_i^{[k]} = V_{e_i}^{[k+1]T}(Ae_i + d_i C_i u_i^{[k+1]} - \sum_{j\in N_i^1} a_{ij} C_j u_j^{[k+1]} - \sum_{j\in N_i^2} a_{ij} C_j u_j^{[k]}) - V_{e_i}^{[k]T}(Ae_i + d_i C_i u_i^{[k]} - \sum_{j\in N_i} a_{ij} C_j u_j^{[k]})$.

According to (8.23), we can obtain

$$
\begin{aligned}
&\dot{V}_i^{[k]} - \dot{V}_i^{[k+1]} \\
&= r_z^i(e_i, u_i^{[k+1]}, u_{N_i^1}^{[k+1]}, u_{N_i^2}^{[k]}) - r_z^i(e_i, u_i^{[k]}, u_{N_i}^{[k]}) \\
&= \sum_{j\in\{N_i^1, i\}} (u_j^{[k+1]} - u_j^{[k]})^T L_{ij} (u_j^{[k+1]} - u_j^{[k]}) \\
&\quad + 2 \sum_{j\in\{N_i^1, i\}} u_j^{[k]T} L_{ij} (u_j^{[k+1]} - u_j^{[k]})
\end{aligned} \tag{8.26}
$$

In order to make $\dot{V}_i^{[k]} - \dot{V}_i^{[k+1]} \geq 0$, the following condition needs to hold:

$$
\triangle u_j^T L_{ij} \triangle u_j \geq -2u_j^{[k]T} L_{ij} \triangle u_j,\ j \in \{N_i^1, i\} \tag{8.27}
$$

where $\triangle u_j = u_j^{[k+1]} - u_j^{[k]}$.

That is to say

$$
\triangle u_j^T L_{ij} \triangle u_j \geq d_j V_{e_j}^{[k-1]T} C_j L_{jj}^{-T} L_{ij} \triangle u_j,\ j \in \{N_i^1, i\} \tag{8.28}
$$

Taking the norm of (8.28), we easily get

$$
\lambda_{min}(L_{ij}) \|\triangle u_j\| \geq d_j \lambda_{max}(L_{jj}^{-T} L_{ij}) \|C_j\| \|V_{e_j}^{[k-1]}\| \tag{8.29}
$$

where $j \in \{N_i^1, i\}$.

Based on the analysis of the above, we obtain $V_i^{[k]} \geq V_i^{[k+1]}$ which implies that it is a nonincreasing function. Meanwhile, it is also bounded below by 0. Thus $V_i^{[k]}$ is convergent, i.e., $\lim\limits_{k\to\infty} V_i^{[k]} = V_i^{\infty}$.

According to (8.14), one has

$$
\begin{aligned}
V_i^{[k]} &\geq \int_t^{\infty} (e_i^T S_{ii} e_i + u_i^{*T} L_{ii} u_i^* + \sum_{j\in N_i} u_j^{*T} L_{ij} u_j^*) d\tau \\
&= V_i^*
\end{aligned} \tag{8.30}
$$

then we have $V_i^{\infty} \geq V_i^*$ when $k \to \infty$.

And we easily obtain $V_i^{\infty} \leq V_i^*$. Thus one has $\lim\limits_{k\to\infty} V_i^{[k]} = V_i^{\infty} = V_i^*$. Then the algorithm converges to V_i^* and it is obtained that $u_i, \forall i$ converges to the Nash Equilibrium.

8.5 Data-driven Approximation Solutions for The Coupled HJB Equations by Critic-Actor Neural Networks

In the section, we will establish the critic-actor network framework based on PI algorithm to solve the coupled HJB equations online, in which the critic neural networks (NNs) will be used to approximate the value functions and the actor NNs can be utilized to perform the control policies.

8.5.1 *Critic Network*

Firstly, we utilize a critic NN to approximate the value function for each agent. Choose $\hat{W}_{ci}(t)$ and $\sigma_{ci}(t)$ as the hidden-to-output layer and input-to-hidden layer weight matrix of the critic network, respectively. Thus the estimated value function $\hat{V}_i(e_i(t))$ of agent i can be formulated as

$$\hat{V}_i(e_i(t)) = \hat{W}_{ci}^T(t)\varphi(\sigma_{ci}^T y_i(t)) \tag{8.31}$$

where $\varphi(\cdot)$ denotes the critic NN activation function, $y_i(t)$ is the generalized input variable including $e_i(t)$ and $u_{\bar{N}_i}(t)$. For convenience of analysis, only the hidden-to-output layer weight matrix $\hat{W}_{ci}(t)$ is updated during the training when the input-to-hidden layer weight matrix $\sigma_{ci} = I$ keeps as constant.

Then the HJ equations (8.23) can be approximated at each step as

$$\begin{aligned} \mathbf{e}_{ci}(t) &= \hat{V}_i(e_i(t+T_s)) - \hat{V}_i(e_i(t)) + \int_t^{t+T_s} r_z^i(e_i, u_{\bar{N}_i})d\tau \\ &= \hat{W}_{ci}^T \Delta\varphi(y_i) + \int_t^{t+T_s} (e_i^T S_{ii} e_i + u_i^T L_{ii} u_i + \sum_{j\in N_i} u_j^T L_{ij} u_j)d\tau \end{aligned} \tag{8.32}$$

where $\Delta\varphi(y_i) = \varphi(y_i(t+T_s)) - \varphi(y_i(t))$.

Given any admissible consensus control policies $u_{\bar{N}_i}$, the desired $\hat{W}_{ci}$ is selected to minimize the squared residual error here as

$$E_{ci}(t) = \frac{1}{2}\mathbf{e}_{ci}^T(t)\mathbf{e}_{ci}(t) \tag{8.33}$$

Based on the gradient descent algorithm [6], the tuning law of the critic NN weight estimation for agent i is presented as

$$\dot{\hat{W}}_{ci}(t) = -\alpha_{ci}\frac{\partial E_{ci}(t)}{\partial \mathbf{e}_{ci}(t)}\frac{\partial \mathbf{e}_{ci}(t)}{\partial \hat{V}_i(e_i(t))}\frac{\partial \hat{V}_i(e_i(t))}{\partial \hat{W}_{ci}(t)}$$

$$= \alpha_{ci}\varphi(y_i(t))(\hat{W}_{ci}^T(t)\Delta\varphi(y_i) + \int_t^{t+T_s} r_z^i(e_i, u_{\bar{N}_i})d\tau)^T \tag{8.34}$$

where $\alpha_{ci} > 0$ is the gain of the critic NN.

8.5.2 Actor Network

Define an actor NN, which is used to approximate the control policy u_i for agent i, as follow:

$$\hat{u}_i(t) = \hat{W}_{ai}^T(t)\phi(\sigma_{ai}^T e_i(t)) \tag{8.35}$$

where $\phi(\cdot)$ denotes the actor NN activation function, $\hat{W}_{ai}(t)$ and $\sigma_{ai}(t)$ is the hidden-output layer and the input-hidden layer weight matrix of the actor NN, respectively. The same as above, we choose the input-hidden layer weight matrix as unit matrix, i.e., $\sigma_{ai} = I$. Thus only the weight matrix of hidden-output layer $\hat{W}_{ai}(t)$ is updated.

In order to find the control action to minimize the function value $\hat{V}_i(e_i(t))$, we define the actor neural network error function as

$$\mathbf{e}_{ai}(t) = \hat{V}_i(e_i(t)) - U_c^i \tag{8.36}$$

where U_c^i is the ultimate utility function. The value of U_c^i is critical in ADP design and it could be variant in different application. In this chapter, we choose $U_c^i = 0$.

The objective function to be minimized by an actor network is defined as

$$E_{ai}(t) = \frac{1}{2}\mathbf{e}_{ai}^T(t)\mathbf{e}_{ai}(t) \tag{8.37}$$

The updating law of actor NN weight estimation is designed as

$$\begin{aligned}
&\dot{\hat{W}}_{ai}(t) \\
&= -\beta_{ai}\frac{\partial E_{ai}(t)}{\partial \mathbf{e}_{ai}(t)}\frac{\partial \mathbf{e}_{ai}(t)}{\partial \hat{V}_i(t)}\frac{\partial \hat{V}_i(t)}{\partial \varphi(y_i(t))}\frac{\partial \varphi(y_i(t))}{\partial y_i(t)}\frac{\partial y_i(t)}{\partial \hat{u}_i(t)}\frac{\partial \hat{u}_i(t)}{\partial \hat{W}_{ai}(t)} \\
&= -\beta_{ai}\phi(e_i(t))\hat{W}_{ci}^T(t)\nabla\varphi(y_i(t))p_i\varphi^T(y_i(t))\hat{W}_{ci}(t)
\end{aligned} \tag{8.38}$$

where $\nabla\varphi(y_i(t)) = \frac{\partial\varphi(y_i(t))}{\partial y_i(t)}$, $p_i = \frac{\partial y_i(t)}{\partial \hat{u}_i(t)}$ is a constant matrix, β_{ai} is the gain of the actor NN, respectively.

Remark 8.2 In the critic-actor NNs, we only utilize the measured data to update the weight of the two NNs rather than the system model matrices A and B_i. Compared with [1, 2, 4] , it is more meaningful in practical applications since the accurate mathematical models of real-world systems are usually difficult to obtain.

8.6 Simulation Example

This section will illustrate the effectiveness of the proposed approach by a simulation example. The MAS under consideration is a connection of four agents as shown in Fig. 8.1. Consider the following continuous-time input-delayed MAS:

$$\dot{x}_i(t) = Ax_i(t) + B_i u_i(t - \tau_i)$$

where $A = \begin{bmatrix} 0.9 & -0.05 \\ -0.75 & 0.1 \end{bmatrix}$, $B_1 = \begin{bmatrix} 0.4 \\ 0 \end{bmatrix}$, $B_2 = \begin{bmatrix} -0.8 \\ 0 \end{bmatrix}$, $B_3 = \begin{bmatrix} 0.2 \\ 0 \end{bmatrix}$, $B_4 = \begin{bmatrix} -0.5 \\ 0 \end{bmatrix}$.

The bounded input delays for four agents are chosen as τ_1=1.5s, $\tau_2 = 0.41s$, $\tau_3 = 0.80s$, $\tau_4 = 0.03s$, the matrices of the performance indices (9.10) are set as $S_{ii} = 2I_2$, $L_{ii} = 5$, $L_{ij} = 1$ and the gain of the critic-actor NNs are $\alpha_{ci} = 3 \times 10^{-4}$, $\beta_{ai} = 5 \times 10^{-4}$. According to the directed communication topology, we select the critic and actor NN activation functions as

$$\begin{aligned}
\varphi_{c1} &= [tanh(e_{11}^2)\ tanh(e_{12}^2)\ tanh(\hat{u}_1^2)\ tanh(\hat{u}_2^2)\ tanh(\hat{u}_4^2)]^T \\
\varphi_{c2} &= [tanh(e_{21}^2)\ tanh(e_{22}^2)\ tanh(\hat{u}_1^2)\ tanh(\hat{u}_2^2)\ tanh(\hat{u}_3^2)]^T \\
\varphi_{c3} &= [tanh(e_{31}^2)\ tanh(e_{32}^2)\ tanh(u_1^2)\ tanh(u_3^2)\ tanh(u_4^2)]^T \\
\varphi_{c4} &= [tanh(e_{41}^2)\ tanh(e_{42}^2)\ tanh(\hat{u}_2^2)\ tanh(\hat{u}_4^2)]^T
\end{aligned}$$

and

$$\phi_i = [tanh(e_{i1})\ tanh(e_{i2})]^T, \quad i = 1, \ldots, 4$$

The evolution of the local neighbor consensus errors for the transformed delay-free MAS is described in Fig. 8.2 which indicates that the transformed delay-free MAS (8.6) can achieve consensus. Figure 8.3 displays the 3-D phase plane plot of the states for the original input-delayed MAS in which all the agents converge to

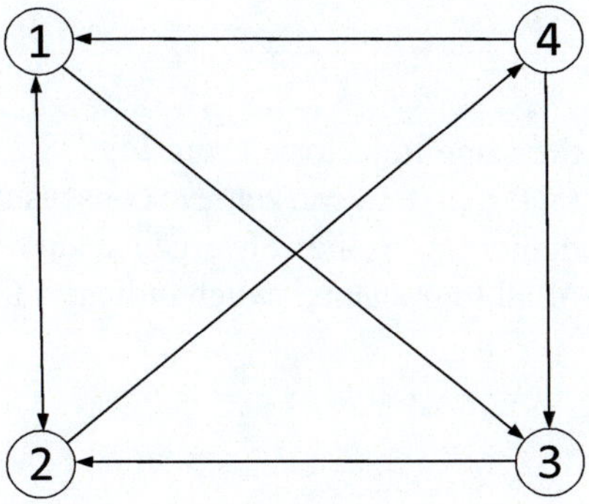

Fig. 8.1 Structure of 4-agent digraph

Fig. 8.2 The local neighbor consensus errors for the transformed delay-free MAS

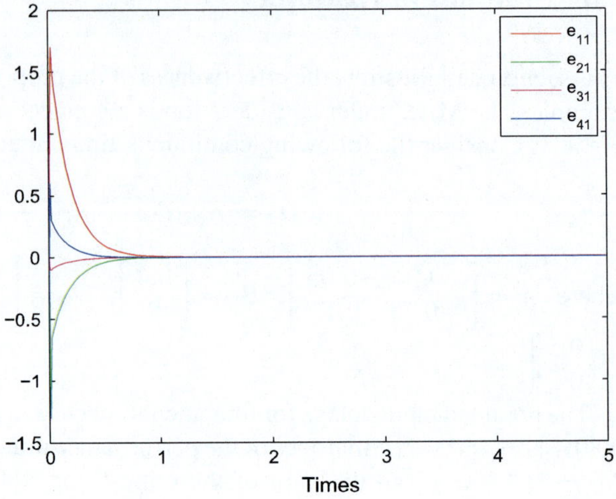

(a)

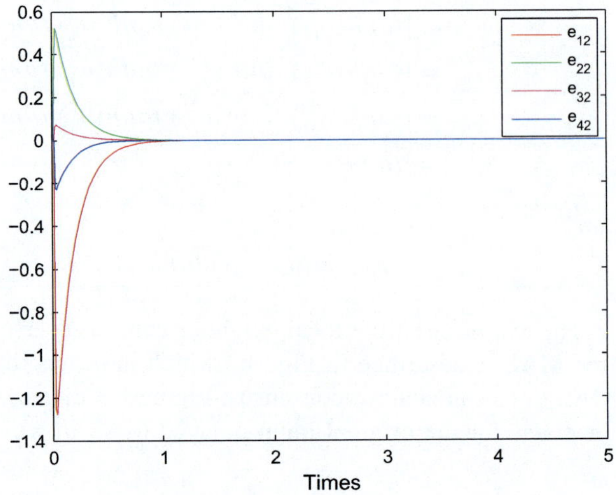

(b)

the same trajectory. From Figs. 8.2 and 8.3, it can easily show that the two MASs (8.1) and (8.6) can achieve consensus under the same controller. Meanwhile, Fig. 8.4 demonstrates the convergence curves of the critic and actor NN weight estimations of all four agents, which indicates that their estimation errors are UUB.

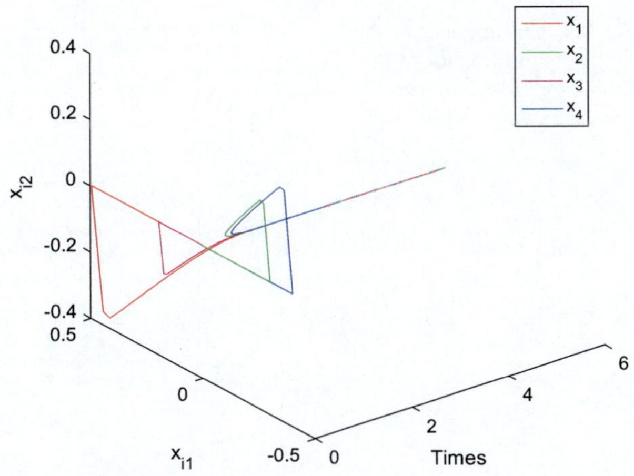

Fig. 8.3 3-D phase plane plot of the states for original input-delayed MAS

A comparison between the proposed optimal consensus controller and the designed consensus controller in [7] is presented in Fig. 8.5. Figure 8.5a shows that the evolution of the control inputs for two method and it implies that the control inputs of our proposed method converge to zero faster than the controller in [7]. Furthermore, Fig. 8.5b shows the performance functions of the proposed method are obviously lower than those of the control scheme in [7]. The simulation results presented in Fig. 8.5 indicate the effectiveness of our proposed optimal consensus controller.

8.7 Conclusion

In this chapter, data-driven distributed optimal consensus control problem for unknown MASs with input delays has been addressed using ADP method. The model reduction method was used to deal with input delays so as to convert the system model into an input-delay-free MAS version. To completely transform optimal consensus control of input-delayed MAS into that of input-delay-free MAS, we established an equivalence relation on the performance indices of the two systems. Then a policy iteration algorithm based on distributed asynchronous update mechanism was proposed to online learn the coupled HJB equations. To preform the proposed adaptive dynamic programming method, we employed data-driven critic-actor NNs to approximate the value functions and the control policies, respectively. Finally, a simulation example was provided to illustrate the effectiveness of the proposed method.

Fig. 8.4 Convergence of the NN weight estimations:(a) Critic NN; (b) Actor NN

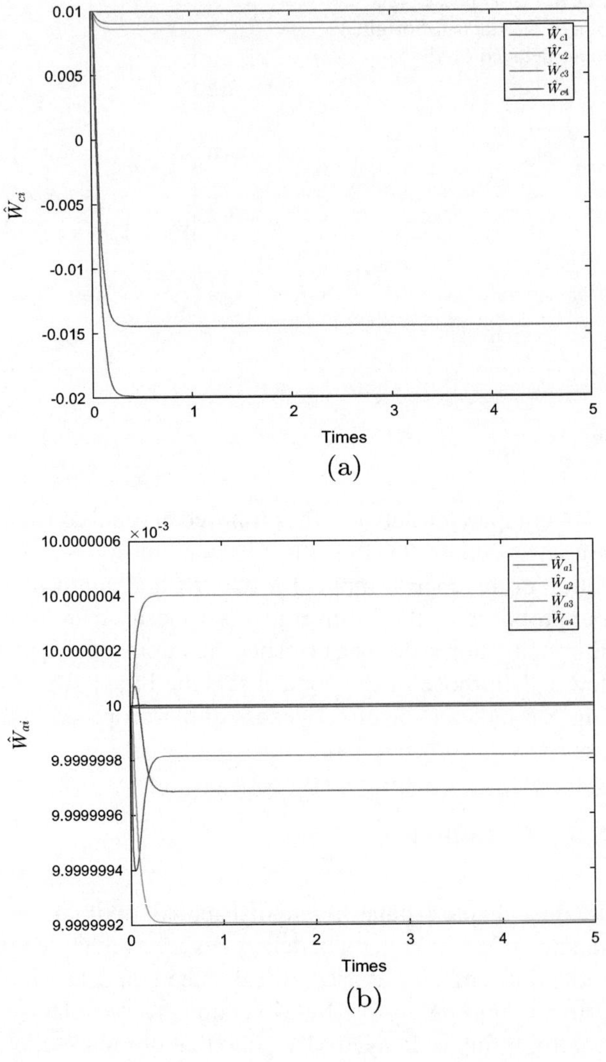

Fig. 8.5 A comparison between the proposed optimal consensus controller and the designed consensus controller in [15] under uniform initial condition: (a) control inputs. (b) performance functions

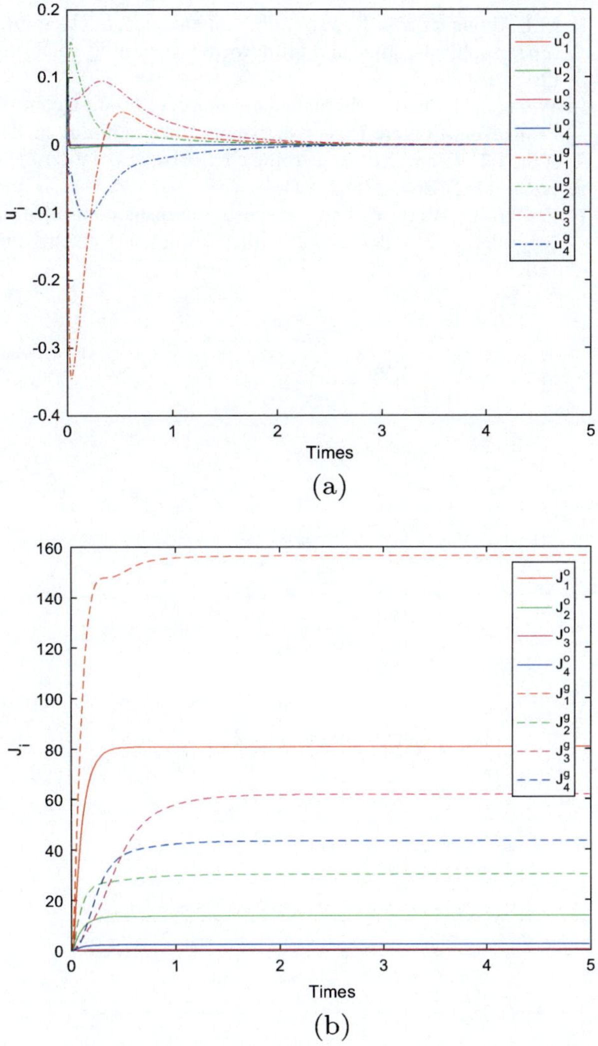

References

1. K.G. Vamvoudakis, F.L. Lewis, G.R. Hudas, Multi-agent differential graphical games: Online adaptive learning solution for synchronization with optimality. Automatica **48**(8), 1598–1611 (2012)
2. H. Zhang, J. Zhang, G.-H. Yang, Y. Luo, Leader-based optimal coordination control for the consensus problem of multiagent differential games via fuzzy adaptive dynamic programming. IEEE Trans. Fuzzy Syst. **23**(1), 152–163 (2015)
3. H. Zhang, H. Jiang, Y. Luo, G. Xiao, Data-driven optimal consensus control for discrete-time multi-agent systems with unknown dynamics using reinforcement learning method. IEEE Trans. Ind. Electron. **64**(5), 4091–4100 (2016)

4. M.I. Abouheaf, F.L. Lewis, K.G. Vamvoudakis, S. Haesaert, R. Babuska, Multi-agent discrete-time graphical games and reinforcement learning solutions. Automatica **50**(12), 3038–3053 (2014)
5. W. Zhao, H. Zhang, Distributed optimal coordination control for nonlinear multi-agent systems using event-triggered adaptive dynamic programming method. ISA Trans. **91**, 184–195 (2019)
6. J. Si, Y.T. Wang, On-line learning control by association and reinforcement. IEEE Trans. Neural Netw. **12**(2), 264–276 (2001)
7. Z. Zuo, C. Wang, Z. Ding, Robust consensus control of uncertain multi-agent systems with input delay: a model reduction method. Int. J. Robust Nonlinear Control **27**(11), 1874–1894 (2017)

Chapter 9
Data-Driven Event-Triggered Optimal Consensus Control for Nonlinear Multiagent Systems

This chapter considers optimal consensus control problem for unknown nonlinear MASs subjected to control constraints by utilizing event-triggered adaptive dynamic programming (ETADP) technique. To deal with the control constraints, we introduce nonquadratic energy consumption functions into performance indices and formulate the Hamilton-Jacobi-Bellman (HJB) equations. Then, based on the Bellman's optimality principle, constrained optimal consensus control policies are designed from the HJB equations. In order to implement the ETADP algorithm, the critic networks and action networks are developed to approximate the value functions and consensus control policies respectively based on the measurable system data. Under the event-triggered control framework, the weights of the critic networks and action networks are only updated at the triggering instants which are decided by the designed adaptive triggered conditions. The Lyapunov method is used to prove that the local neighbor consensus errors and the weight estimation errors of the critic networks and action networks are ultimately bounded. Finally, a numerical example is provided to show the effectiveness of the proposed ETADP method.

Optimal event-triggered consensus control policies are designed in Sect. 9.1. To implement the algorithm, we use the gradient descent method to design the weight updated laws of the critic-action neural networks in Sect. 9.2. The stability analysis is provided in Sect. 9.3. Before concluding in Sect. 9.5, an example is illustrated to show the effectiveness of the proposed method in Sect. 9.4.

9.1 Problem Formulation

Consider a nonlinear MAS consisting of N agents, labeled from v_1 to v_N. The dynamics of agent i are represented as

$$x_i(k+1) = f(x_i(k)) + g_i(x_i(k))u_i(k) \tag{9.1}$$

where $x_i(k) \in R^n$ and $u_i(k) \in R^m$ denote the state and control input, $f(\cdot)$ and $g_i(\cdot)$ are the unknown drift dynamics and the unknown control effectiveness. Denote

D. Yue et al., *Distributed Cooperative Control and Communication for Multi-agent Systems*, https://doi.org/10.1007/978-981-33-6718-0_9

$\Omega_u = \{u_i(k) | u_i^T(k)u_i(k) \leq \Lambda\}$ as the set of all control inputs satisfying the control constraints, in which Λ is the constraint bound for the actuator of each agent.

Assumption 9.1 $f + g_i u_i$ is Lipschitz continuous and $f(0) = 0$.

Assumption 9.2 The control effectiveness $g_i(\cdot)$ is bounded such that $\|g_i(\cdot)\| \leq g_M$ with g_M being a known positive constant.

To consider consensus problem for MAS (9.1), we define the neighbor consensus error of agent i as

$$\delta_i(k) = \sum_{j \in N_i} a_{ij}(x_i(k) - x_j(k)) = \sum_{j \in \bar{N}_i} l_{ij} x_j(k). \tag{9.2}$$

Subsequently for agent i, the dynamics of the neighbor consensus error are calculated as

$$\begin{aligned} \delta_i(k+1) &= \sum_{j \in \bar{N}_i} l_{ij} x_j(k+1) \\ &= \sum_{j \in \bar{N}_i} l_{ij}(f(x_j(k)) + g_j(x_j(k))u_j(k)). \end{aligned} \tag{9.3}$$

Then the local value function for agent i is defined as

$$\begin{aligned} V_i(k) &= \sum_{l=k}^{\infty} U_i(\delta_i(l), u_{\bar{N}_i}(l)) \\ &= \sum_{l=k}^{\infty} \{\delta_i^T(l) Q_i \delta_i(l) + \sum_{j \in \bar{N}_i} W_j(u_j(l))\} \end{aligned} \tag{9.4}$$

where $U_i(\delta_i(l), u_{\bar{N}_i}(l)) = \delta_i^T(l) Q_i \delta_i(l) + \sum_{j \in \bar{N}_i} W(u_j(l))$ denotes the cost-to-go function, matrix Q_i and function $W_j(u_j(l))$ are positive definite.

In existing works on unconstrained optimal consensus control [1–3], $W_i(u_i)$ is generally designed as the quadratic function of the control policy u_i, that is, $W_i(u_i) = u_i^T S_i u_i$ with S_i being a positive definite matrix. When confronting optimal consensus problem with control constraints, we select a non-quadratic function as

$$\begin{aligned} W_i(u_i(l)) &= 2 \int_0^{u_i(l)} \Lambda \varphi^{-T}(\nu/\Lambda) R_i d\nu, \quad i = 1, 2, \ldots, N \\ \varphi^{-T}(\nu/\Lambda) &= [\varphi^{-1}(\nu_1/\Lambda), \varphi^{-1}(\nu_2/\Lambda), \ldots, \varphi^{-1}(\nu_m/\Lambda)]^T \end{aligned} \tag{9.5}$$

for agent i, where R_i is positive definite diagonal matrix, which is represented as $R_i = diag(r_{i1}, \ldots, r_{im})$. $\varphi(\cdot) \in \mathbb{R}^m$ is a bounded one-to-one function which satisfies

$\|\varphi(\cdot)\| \leq 1$ and belong to $C^p(p \geq 1)$ and $L_2(\Omega)$. In addition, it is also a monotonic odd function with its first derivative bounded. In general, we select hyperbolic tangent function $\varphi(\cdot) = \tanh(\cdot)$.

Combined with (9.4), we can have

$$\begin{aligned} V_i(k) &= \sum_{l=k}^{\infty} \left\{ \delta_i^T(l) Q_i \delta_i(l) + \sum_{j \in \bar{N}_i} W_j(u_j(l)) \right\} \\ &= \delta_i^T(k) Q_i \delta_i(k) + \sum_{j \in \bar{N}_i} W_j(u_j(k)) + \sum_{l=k+1}^{\infty} \left\{ \delta_i^T(l) Q_i \delta_i(l) + \sum_{j \in \bar{N}_i} W_j(u_j(l)) \right\} \\ &= \delta_i^T(k) Q_i \delta_i(k) + \sum_{j \in \bar{N}_i} W_j(u_j(k)) + V_i(k+1). \end{aligned} \tag{9.6}$$

Based on Bellman's principle of optimality, we can obtain the coupled HJB equation:

$$V_i^*(k) = \min_{u_i(k)} \left\{ \delta_i^T(k) Q_i \delta_i(k) + \sum_{j \in \bar{N}_i} W_j(u_j(k)) + V_i^*(k+1) \right\} \tag{9.7}$$

for agent i where $V_i^*(k)$ denotes optimal value function for agent i.

To Solve the first order necessary condition $\partial V_i^*(k)/\partial u_i(k) = 0$, the constrained optimal consensus controller is derived as

$$u_i^*(k) = -\Lambda \varphi\left(\frac{d_i}{2\Lambda} R_i^{-1} g_i^T(x_i(k)) V_{\delta_i}^*(k+1) \right) = -\Lambda \varphi(p_i) \tag{9.8}$$

where $p_i \triangleq \frac{d_i}{2\Lambda} R_i^{-1} g_i^T(x_i(k)) V_{\delta_i}^*(k+1)$ and $V_{\delta_i}^*(k+1) = \frac{\partial V_i^*(k+1)}{\partial \delta_i(k+1)}$.

9.2 Event-Triggered Optimal Consensus Control Design

In the traditional ADP method [4–6], the system's information are sampled and transmitted in a fix time period. To save resource utilization and reduce transmission burden, the ETADP, which combines event-triggered control with adaptive dynamic programming, was introduced in [7, 8]. A key point to ETADP is to build an adaptive triggered mechanism which determines when an optimal control task is executed.

In order to characterize the event-triggered sampling, suppose that the sequence of the triggering instants is $\{k_s^i\}_{s=0}^{\infty}$ satisfying $0 \leq k_0^i < k_1^i < k_2^i < \cdots < k_s^i < \cdots$, where k_s^i represents sth consecutive event-triggered sampling instant for agent i. Then, for $k \in [k_s^i, k_{s+1}^i)$, the triggering error vector for agent i is defined as

$$e_i(k) = \bar{\delta}_i - \delta_i(k) \tag{9.9}$$

for agent i where $\bar{\delta}_i = \delta_i(k_s^i)$ is the neighbor consensus error at the latest triggering instant $k = k_s^i$. Note that the triggering error $e_i(k)$ is reset to zero at triggered instants, i.e., $e_i(k_s^i) = 0$, $s \in N_+$. Then, the sequence of triggered instants is determined iteratively by

$$k_{s+1}^i = \{k \in \mathbb{N}^+ | k > k_s^i, \Upsilon_i(e_i(k), \xi_i) = e_i^T(k)e_i(k) - \xi_i > 0\} \tag{9.10}$$

where $\Upsilon_i(\cdot, \cdot) : \mathbb{R}^n \times \mathbb{R} \longrightarrow \mathbb{R}$ denotes the event-triggering condition of agent i, ξ_i denotes the threshold function for agent i under the event-triggered scheme which depends on the information of agent i and its neighbors.

Under event-triggered scheme, the optimal event-triggered consensus controller of agent i can be written as

$$\mu_i = -\Lambda\varphi\left(\frac{d_i}{2\Lambda}R^{-1}g_i^T(\bar{x}_i)V_{\bar{\delta}_i}^*\right) = -\Lambda\varphi(\bar{p}_i), k = k_s^i \tag{9.11}$$

where $\bar{p}_i \triangleq \frac{d_i}{2\Lambda}R^{-1}g_i^T(\bar{x}_i)V_{\bar{\delta}_i}^*$ and $\bar{x}_i = x_i(k_s^i)$ is the state at the triggering instant.

Substituting (9.11) into (9.7), it yields the coupled HJB equations for event-triggered nonlinear MASs with control constraints as

$$V_i^*(k) = \min_{u_i(k)}\{\delta_i^T(k)Q_i\delta_i(k) + \sum_{j\in\bar{N}_i} 2\int_0^{-\Lambda\varphi(\bar{p}_j)} \Lambda\varphi^{-T}(\nu/\Lambda)R_j d\nu + V_i^*(k+1)\}. \tag{9.12}$$

Due to the coupled HJB equations (9.12) is nonlinear, it is difficult to obtain the analytical solutions. To learn the solutions to the HJB equations online, value or policy iteration-based ADP method was proposed in [1, 2, 9]. Since the number of iterations needed to obtain suitable value function and control policy approximations is unknown, it is generally not suitable for online learning. So a novel adaptive method without value iteration which online approximates the value functions and consensus control policies using neural network structure is proposed in Section 4. Here, the value functions and control policies are updated aperiodically so as to make them highly suitable for online learning.

9.3 Online Solution of Optimal Consensus Controllers Using Neural Networks

In the section, a novel ETADP scheme is firstly for each agent designed which includes event generator (EG), smart critic network, and smart action network. Then, the critic-action networks are built to approximate the value functions and control policies, respectively. Subsequently, the stability analysis of the proposed method is provided.

9.3.1 ETADP Approach

The new structure of ETADP method based on the measurable system data is designed in Fig. 9.1. For each agent, the proposed scheme includes two networks where a smart critic network consisting of event generator and critic network is built to approximate the value function $V_i(\delta_i)$, and a smart action network consisting of event generator and action network is utilized to learn the optimal constrained consensus control policy μ_i.

In Fig. 9.1, EG is built to calculate the triggering error (9.9) and to evaluate the triggering condition (9.10) at each time instant. When the designed triggering condition (9.10), which depends on the triggering error, the system data, and the estimated action NN weight, is violated, the current time instant and current state are recorded as a sampled instant and sampled state, respectively. At this triggering instant, meanwhile, the current sampled state is transmitted to its neighbor agents N_i and the current local neighbor consensus error is sent to the critic-action networks. To simplify the calculation and training process, the triggering mechanisms of smart critic-action networks are chosen to be same with that of EG. Thus, the wights of the critic-action networks are only updated at the triggering instants and hold constant during the inter-triggered time intervals.

Remark 9.1 Compared with the optimal ETC system in [10], the proposed ETADP method which has the same triggered mechanism in EG and smart critic-action networks can reduce the amount of calculation and it is easy to implement in practical systems.

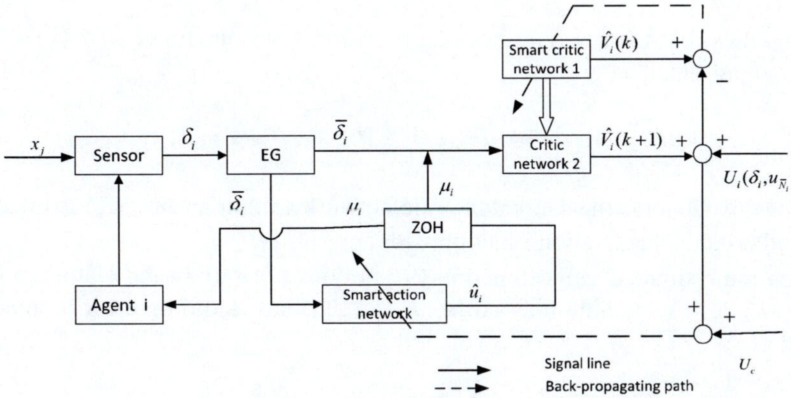

Fig. 9.1 Schematic diagram of ETADP approach for agent *i*

9.3.2 Smart Critic-Action Neural Networks Design

Here, we focus on the optimal consensus control policies design by using the proposed ETADP approach, which are implemented by employing the critic-action network structures to solve the event-based HJB equations (9.12). For each agent, a critic network is used to approximate the value function while an action network is employed for control policy approximation.

(1) Critic NN design
According to the universal approximation property of neural network (NN) [11], the critic NN is used to approximate the value function which can be formulated as

$$V_i(k) = W_{ci}^T \psi_{ci}(z_i(k)) + \varepsilon_{ci}(k) \tag{9.13}$$

for agent i where $W_{ci} \in \mathbb{R}^{h_{v_i}}$ is the ideal weight of the critic network with h_{v_i} being the number of hidden layer neurons, $\psi_{ci}(\cdot) \in \mathbb{R}^{h_{v_i}}$ is time-varying activation function of the critic network, $z_i(k) = [\delta_i^T(k), u_{\bar{N}_i}^T(k)]^T$ denotes the generalized input variable, $\varepsilon_{ci}(k)$ is the reconstruction error of the critic network, and it satisfies $\|\varepsilon_{ci}\| \leq \varepsilon_{c,M}$ with $\varepsilon_{c,M} > 0$.

From (9.6) and (9.13), the Hamilton equations can be rewritten as

$$W_{ci}^T \Delta\psi_{ci}(z_i(k)) + \delta_i^T(k) Q_i \delta_i(k) + \sum_{j \in \bar{N}_i} W_j(u_j(k)) + \Delta\varepsilon_{ci}(k) = 0 \tag{9.14}$$

where $\Delta\psi_{ci}(z_i(k)) = \psi_{ci}(z_i(k+1)) - \psi_{ci}(z_i(k))$ and $\Delta\varepsilon_{ci}(k) = \varepsilon_{ci}(k+1) - \varepsilon_{ci}(k)$.

Under the event-triggered framework, the estimated value function $\hat{V}_i(k)$ of agent i can be represented as

$$\hat{V}_i(k) = \hat{W}_{ci}^T(k) \psi_{ci}(\bar{z}_i), \quad k \in [k_s^i, k_{s+1}^i), \quad s = 1, 2, \ldots \tag{9.15}$$

where $\hat{W}_{ci}(k)$ denotes the estimated critic network weight and $\psi_{ci}(\bar{z}_i)$ is the event-triggered critic NN activation function with $\bar{z}_i = [\bar{\delta}_i^T, \mu_{\bar{N}_i}^T]^T$.

Since the estimated value function (9.15) does not satisfy the Hamilton equation (9.14) during the time interval $[k_s^i, k_{s+1}^i)$, thus the Hamilton error of agent i is defined as

$$\begin{aligned} e_{ci}(k) &= \delta_i^T(k) Q_i \delta_i(k) + \sum_{j \in \bar{N}_i} W_j(u_j(k)) + \hat{V}_i(k+1) - \hat{V}_i(k) \\ &= \hat{W}_{ci}^T(k) \Delta\psi_{ci}(\bar{z}_i) + \delta_i^T(k) Q_i \delta_i(k) + \sum_{j \in \bar{N}_i} W_j(u_j(k)). \end{aligned} \tag{9.16}$$

Meanwhile, at the triggered instants $k = k_s^i$, $s = 1, 2, \ldots$, the Hamilton error of agent i can be written as

$$\begin{aligned} e_{ci}(k) &= \bar{\delta}_i^T Q_i \bar{\delta}_i + \sum_{j \in \bar{N}_i} W_j(\mu_j) + \hat{V}_i(k_s^i + 1) - \hat{V}_i(k_s^i) \\ &= \hat{W}_{ci}^T \Delta\psi_{ci}(\bar{z}_i) + \bar{\delta}_i^T Q_i \bar{\delta}_i + \sum_{j \in \bar{N}_i} W_j(\mu_j) \end{aligned} \tag{9.17}$$

where $\hat{V}_i(k_s^i + 1) = \hat{W}_{ci}^T(k_s^i)\psi_{ci}(z_i(k_s^i + 1))$ and $\Delta\psi_{ci}(\bar{z}_i) = \psi_{ci}(z_i(k_s^i + 1)) - \psi_{ci}(\bar{z}_i)$.

The objective of the critic network for agent i is to minimize the error performance index

$$E_{ci}(k) = \frac{1}{2} e_{ci}^T(k) e_{ci}(k). \tag{9.18}$$

In the event-triggered framework, the weight update law of the critic network for agent i is selected as

$$\hat{W}_{ci}(k) = \begin{cases} \hat{W}_{ci}(k-1) - \gamma_{ci} \frac{\Delta\psi_{ci}(z_i(k-1)) e_{ci}^T(k-1)}{\Delta\psi_{ci}^T(z_i(k-1))\Delta\psi_{ci}(z_i(k-1))+1}, & k = k_s^i; \\ \hat{W}_{ci}(k-1), \quad k \in (k_s^i, k_{s+1}^i). \end{cases} \tag{9.19}$$

where $\gamma_{ci} > 0$ is the learning gain.

Remark 9.2 The weights of the critic networks are only tuned at the triggering instants and held invariability during the inter-event times. This updated scheme saves the number of computation and reduces transmission load compared with the time-triggered ADP method [2, 3].

Subsequently, we define the critic network weight estimation error for agent i as

$$\tilde{W}_{ci}(k) = W_{ci} - \hat{W}_{ci}(k). \tag{9.20}$$

Then based on (9.19) and (9.20), we can obtain

$$\tilde{W}_{ci}^T(k+1) = \begin{cases} \tilde{W}_{ci}(k) + \gamma_{ci} \frac{\Delta\psi_{ci}(z_i(k)) e_{ci}^T(k)}{\Delta\psi_{ci}^T(z_i(k))\Delta\psi_{ci}(z_i(k))+1}, & k = k_s^i; \\ \tilde{W}_{ci}(k), & k \in (k_s^i, k_{s+1}^i). \end{cases} \tag{9.21}$$

Adding (9.6) into (9.16), one has

$$e_{ci}(k) = W_{ci}^T(\Delta\psi(\bar{z}_i) - \Delta\psi(z_i(k))) - \tilde{W}_{ci}^T \Delta\psi(z_i(k)) + \bar{\delta}_i^T Q_i \bar{\delta}_i \tag{9.22}$$

$$- \delta_i^T Q_i \delta_i - \Delta\varepsilon_{ci}(k) \quad k \in (k_s^i, k_{s+1}^i). \tag{9.23}$$

Moreover, at the triggering instants $k = k_s^i$, (9.22) can be expressed as

$$e_{ci}(k) = -\tilde{W}_{ci}^T \Delta\psi(z_i(k)) - \Delta\varepsilon_{ci}(k) \qquad k = k_s^i. \tag{9.24}$$

(2) Action NNs design
Similar to the approximation of the value function, we employ the action NN to approximate consensus control policy, which is represented as

$$u_i(k) = W_{ai}^T \psi_{ai}(z_i(k)) + \varepsilon_{ai}(k) \tag{9.25}$$

for agent i where $W_{ai} \in \mathbb{R}^{h_{u_i} \times m}$ is the ideal action NN weight from the hidden layer to the output layer in which h_{u_i} is the number of hidden layer neurons, $\psi_{ai}(\cdot) \in \mathbb{R}^{h_{u_i}}$ is the activation function for the action network, and $\varepsilon_{ai}(k)$ is the reconstruction error and it satisfies $\|\varepsilon_{ai}\| \le \varepsilon_{a,M}$ with $\varepsilon_{a,M} > 0$.

Let $\hat{W}_{ai}(k)$ be the action NN weight estimation for agent i, then the estimated consensus control policy in the event-triggered context can be represented as

$$\hat{\mu}_i(k) = \hat{W}_{ai}^T(k)\psi_{ai}(\bar{z}_i), \quad k \in [k_s^i, k_{s+1}^i), \quad s = 1, 2, \ldots. \tag{9.26}$$

For agent i, the action NN error function is defined as

$$e_{ai}(k) = \hat{V}_i(\bar{\delta}_i) - U_c \tag{9.27}$$

where U_c denotes the desired ultimate objective and it is set to 0.

The objective of the action network is to minimize the error performance index

$$E_{ai}(k) = \frac{1}{2} e_{ai}^T(k) e_{ai}(k). \tag{9.28}$$

Based on the gradient descent algorithm, the weight update law of the critic network is selected as

$$\hat{W}_{ai}^T(k) = \begin{cases} \hat{W}_{ai}^T(k-1) - \gamma_{ai}\psi_{ai}(\bar{z}_i)\hat{W}_{ci}^T(k-1)\psi_{ci}'(\bar{z}_i)p_i[\hat{W}_{ci}^T(k-1)\psi_{ci}(\bar{z}_i)]^T, & k = k_s^i; \\ \hat{W}_{ai}^T(k-1), & k \in (k_s^i, k_{s+1}^i). \end{cases} \tag{9.29}$$

where $\psi_{ci}'(\bar{z}_i) = \frac{\partial \psi_{ci}(z_i(k))}{\partial z_i(k)}|_{z_i(k)=\bar{z}_i}$, $p_i = \frac{\partial z_i(k)}{\partial \hat{\mu}_i(k)}$ and $\gamma_{ai} > 0$ is the learning gain.

Define the action NN weight estimation error for agent i as $\tilde{W}_{ai}(k) = W_{ai} - \hat{W}_{ai}(k)$. Based on (9.29), the dynamics of $\tilde{W}_{ai}(k)$, moving one time step, are presented as

$$\tilde{W}_{ai}^T(k+1) = \begin{cases} \tilde{W}_{ai}(k) - \gamma_{ai}\psi_{ai}(\bar{z}_i)\hat{W}_{ci}^T(k)\psi_{ci}'(\bar{z}_i)p_i[\hat{W}_{ci}^T(k)\psi_{ci}(\bar{z}_i)]^T, & k = k_s^i; \\ \tilde{W}_{ai}(k), & k \in (k_s^i, k_{s+1}^i). \end{cases} \tag{9.30}$$

Remark 9.3 The weight update laws of the critic-action networks (9.19) and (9.29) depend on the past and current measurable system's states and control inputs rather than the system functions f_i and g_i which can eliminate the accumulative residual errors caused by model identification method [8, 10]. Compared with optimal coordination control in [1, 2], the proposed method doesnt build policy or value iteration algorithms. What's more, it is more promising in engineering applications since the accurate system dynamic models for MASs are generally difficult or impossible to build.

9.3.3 Stability Analysis

This section analyzes the stability of the proposed algorithm and designs adaptive event-triggered conditions. To analyze the stability of the neighbor consensus error systems (9.2), the following lemma and assumption is introduced.

Lemma 9.1 ([12]) *Consider the nonlinear MAS* (9.1). *There exists optimal consensus control policy* $u_i^*(k)$ *such that the closed-loop dynamics satisfy the inequality*

$$\|f(x_i(k)) + g_i(x_i(k))u_i(k)\|^2 \leq K\|x_i(k)\|^2, \forall i \tag{9.31}$$

where $0 < K < 1$ *is a constant.*

Assumption 9.3 (1) The target weights of the critic-action NNs are bounded, i.e., $\|W_{ci}\| \leq W_{c,M}$ and $\|W_{ai}\| \leq W_{a,M}$ where $W_{c,M}$ and $W_{a,M}$ are positive constants. (2) The critic NN activation functions and their derivative are bounded, i.e., $\|\psi_{ci}\| \leq \psi_{c,M}$ and $\|\partial\psi_{ci}(\cdot)/\partial(\cdot)\| \leq \|\nabla\psi_{c,M}\|$ with $\psi_{c,M}$ and $\nabla\psi_{c,M}$ being positive constants. (3) The activation functions of the action NNs are Lipschitz continuous, i.e., $\|\psi_{ai}(z_i(k)) - \psi_{ai}(\bar{z}_i)\| \leq C\|e_i(k)\|$, where C is a positive constant.

Next, we will show a theorem which designs the adaptive event-triggered condition and considers the ultimate boundedness of the weight estimation error for the critic-action networks.

Theorem 9.1 *Consider the nonlinear MAS* (9.1) *with control constraint and let Assumption* 9.1 *though Assumption* 9.3 *hold. The weights of the critic-action networks are tuned as* (9.19) *and* (9.29), *respectively. Then, the neighbor consensus error* $\delta_i(k)$ *and the weight estimation errors for the critic-action networks* $\tilde{W}_{ci}(k)$, $\tilde{W}_{ai}(k)$ *are ultimately bounded (UB) under the event-triggered conditions*

$$\|e_i(k)\| \leq \sqrt{\rho_i \frac{(1-2K)\|\delta_i(k)\|^2}{4\sum_{j\in\bar{N}_i} l_{ij}^2 g_M^2 C^2 \|\hat{W}_{aj}(k)\|^2}} \tag{9.32}$$

where $K \in (0, 1/2)$ *and* $\rho_i \in (0, 1)$ *is the designed parameter.*

Proof Consider the candidate Lyapunov function as

$$L_i(k) = L_{\delta i}(k) + L_{vi}(k) + L_{ui}(k) \tag{9.33}$$

where $L_{\delta i}(k) = \delta_i^T(k)\delta_i(k)$, $L_{vi}(k) = \frac{1}{\gamma_{ci}} tr(\tilde{W}_{ci}^T(k)\tilde{W}_{ci}(k))$ and $L_{ui} = \frac{1}{\gamma_{ai}} tr(\tilde{W}_{ai}^T(k)$ $\tilde{W}_{ai}(k))$.

To analyze the system stability, the following two cases are considered.

Case 1: Inter-event time interval, i.e., $k \in (k_s^i, k_{s+1}^i)$.
According to the actual consensus control policy (9.25) and the approximated consensus control policy (9.26), the neighbor consensus error dynamics for agent i during the time interval $[k_s^i, k_{s+1}^i)$ can be represented as

$$\begin{aligned}\delta_i(k+1) = &\sum_{j\in\bar{N}_i} l_{ij}(f(x_j(k)) + g_j(x_j(k))u_j(k) - g_j(x_j(k))(\tilde{W}_{aj}^T(k)\psi_{aj}(z_j(k)) + \varepsilon_{aj}(k)) \\ &- g_j(x_j(k))\hat{W}_{aj}^T(k)(\psi_{aj}(z_j(k)) - \psi_{aj}(\bar{z}_j))).\end{aligned} \tag{9.34}$$

Taking the first difference of $L_{\delta i}(k)$ along the neighbor consensus error dynamics (9.34) yields

$$\begin{aligned}\triangle L_{\delta i}(k) &= \delta_i^T(k+1)\delta_i(k+1) - \delta_i^T(k)\delta_i(k) \\ &= \| \sum_{j\in\bar{N}_i} l_{ij}(f(x_j(k)) + g_j(x_j(k))u_j(k) - g_j(x_j(k))(\tilde{W}_{aj}^T(k)\psi_{aj}(z_j(k)) + \varepsilon_{aj}(k)) \\ &- g_j(x_j(k))\hat{W}_{aj}^T(k)(\psi_{aj}(z_j(k)) - \psi_{aj}(\bar{z}_j)))\|^2 - \| \sum_{j\in\bar{N}_i} l_{ij}x_j(k)\|^2 \\ &\le 2\| \sum_{j\in\bar{N}_i} l_{ij}(f(x_j(k)) + g_j(x_j(k))u_j(k))\|^2 + 4\| \sum_{j\in\bar{N}_i} l_{ij}g_j(x_j(k))(\tilde{W}_{aj}^T(k)\psi_{aj}(z_j(k)) + \varepsilon_{aj}(k))\|^2 \\ &+ 4\| \sum_{j\in\bar{N}_i} l_{ij}g_j(x_j(k))\hat{W}_{aj}^T(k)(\psi_{aj}(z_j(k)) - \psi_{aj}(\bar{z}_j)))\|^2 - \| \sum_{j\in\bar{N}_i} l_{ij}x_j(k)\|^2.\end{aligned} \tag{9.35}$$

Combined Lemma 9.1 with using the equation $(\sum_{p=1}^r a_i y_i)^2 \le r\sum_{p=1}^r a_i^2 y_i^2$, the first difference of $L_{\delta i}(k)$ becomes

$$\begin{aligned}\triangle L_{\delta i}(k) \le &-(1-2K)s_i\Big(\sum_{j\in\bar{N}_i} l_{ij}^2\|x_j(k)\|^2\Big) + \sum_{j\in\bar{N}_i} 8s_i l_{ij}^2 g_M^2(\psi_{a,M}^2\|\tilde{W}_{aj}(k)\|^2 + \varepsilon_{a,M}^2) \\ &+ \sum_{j\in\bar{N}_i} 4s_i l_{ij}^2 g_M^2\|\hat{W}_{aj}^T(k)(\psi_{aj}(z_j(k)) - \psi_{aj}(\bar{z}_j(k)))\|^2.\end{aligned} \tag{9.36}$$

where $s_i = \|\bar{N}_i\|$ denotes the number of agent $\bar{N}_i$.

Based on the Lipschitz continuity of activation function of the action network in Assumption 9.3, we can obtain

$$\triangle L_{\delta i}(k) \leq -(1-2K)s_i\Big(\sum_{j\in\bar{N}_i} l_{ij}^2\|x_j(k)\|^2\Big) + \sum_{j\in\bar{N}_i} 8s_i l_{ij}^2 g_M^2(\psi_{a,M}^2\|\tilde{W}_{aj}(k)\|^2 + \varepsilon_{a,M}^2)$$
$$+ \sum_{j\in\bar{N}_i} 4s_i l_{ij}^2 g_M^2 C^2\|\hat{W}_{aj}(k)\|^2\|e_j(k)\|^2. \tag{9.37}$$

Recalling the event-triggered condition (9.32), it holds that

$$\begin{aligned}\triangle L_{\delta i}(k) &\leq -(1-2K)(1-\rho_i)s_i\Big(\sum_{j\in\bar{N}_i} l_{ij}^2\|x_j(k)\|^2\Big) + \sum_{j\in\bar{N}_i} 8s_i l_{ij}^2 g_M^2(\psi_{a,M}^2\|\tilde{W}_{aj}(k)\|^2 + \varepsilon_{a,M}^2)\\ &\leq -(1-2K)(1-\rho_i)\|\sum_{j\in\bar{N}_i} l_{ij}x_j(k)\|^2 + \sum_{j\in\bar{N}_i} 8s_i l_{ij}^2 g_M^2(\psi_{a,M}^2\|\tilde{W}_{aj}\|^2 + \varepsilon_{a,M}^2)\\ &\leq -(1-2K)(1-\rho_i)\|\delta_i(k)\|^2 + \sum_{j\in\bar{N}_i} 8s_i l_{ij}^2 g_M^2(\psi_{a,M}^2\|\tilde{W}_{aj}\|^2 + \varepsilon_{a,M}^2).\end{aligned} \tag{9.38}$$

where $\rho_i \in (0, 1)$ and $K \in (0, 1/2)$.

Considering the remaining terms of $L_i(k)$, the first differences of them become zero due to the weights of the critic-action NNs are not updated. That is to say,

$$\triangle L_{vi}(k) = 0, \quad \triangle L_{ui}(k) = 0. \tag{9.39}$$

Combined (9.38) and (9.39), the first difference of $L_i(k)$ is calculated as

$$\triangle L_i(k) \leq -(1-2K)(1-\rho_i)\|\delta_i(k)\|^2 + \sum_{j\in\bar{N}_i} 8s_i l_{ij}^2 g_M^2(\psi_{a,M}^2\|\tilde{W}_{aj}\|^2 + \varepsilon_{a,M}^2).$$

In order to guarantee $\triangle L_i(k) < 0$, one must satisfy

$$\|\delta_i(k)\| \geq \sqrt{\Big(\sum_{j\in\bar{N}_i} 8s_i l_{ij}^2 g_M^2(\psi_{a,M}^2\|\tilde{W}_{aj}\|^2 + \varepsilon_{a,M}^2)\Big)/(1-2K)(1-\rho_i)}.$$

Thus, the local neighbor consensus errors are UB. The weight estimation errors of the critic-action networks hold constant during the interval duration so they are also UB.

Case 2: Event-triggering instants, i.e., $k = k_s^i$.
According to (9.34), the neighbor error dynamics at the triggering instants $k = k_s^i$ become

$$\delta_i(k+1) = \sum_{j\in\bar{N}_i} l_{ij}(f(x_j(k)) + g_j(x_j(k))u_j(k) - g_j(x_j(k))(\tilde{W}_{aj}^T(k)\psi_{aj}(z_j(k)) + \varepsilon_{aj}(k))). \tag{9.40}$$

The first difference of $L_{\delta i}(k)$ along (9.40) with Lemma 1 can be written as

$$
\begin{aligned}
&\triangle L_{\delta i}(k)\\
&= \delta_i^T(k+1)\delta_i(k+1) - \delta_i^T(k)\delta_i(k)\\
&= \Big\| \sum_{j\in\bar{N}_i} l_{ij}(f(x_j(k)) + g_j(x_j(k))u_j(k) - g_j(x_j(k))(\tilde{W}_{aj}^T\psi_{aj}(z_j(k)) + \varepsilon_{aj}(k)))\Big\| - \Big\| \sum_{j\in\bar{N}_i} l_{ij}x_j(k)\Big\|^2\\
&\le 2\Big\| \sum_{j\in\bar{N}_i} l_{ij}(f(x_j(k)) + g_j(x_j(k))u_j(k))\Big\|^2 + 2\Big\| \sum_{j\in\bar{N}_i} l_{ij}g_j(x_j(k))(\tilde{W}_{aj}^T\psi_{aj}(z_j(k))\\
&+ \varepsilon_{aj}(k))\Big\|^2 - \Big\| \sum_{j\in\bar{N}_i} l_{ij}x_j(k)\Big\|^2\\
&\le -(1-2K)s_i\Big(\sum_{j\in\bar{N}_i} l_{ij}^2\|x_j(k)\|^2\Big) + \sum_{j\in\bar{N}_i} 2s_i l_{ij}^2 g_M^2(\psi_{a,M}^2\|\tilde{W}_{aj}\|^2 + \varepsilon_{a,M}^2)\\
&\le -(1-2K)\|\delta_i(k)\|^2 + \sum_{j\in\bar{N}_i} 2s_i l_{ij}^2 g_M^2(\psi_{a,M}^2\|\tilde{W}_{aj}\|^2 + \varepsilon_{a,M}^2).
\end{aligned}
\tag{9.41}
$$

Then, consider the second term of $L_{vi}(k)$. The first difference can be presented as

$$
\begin{aligned}
&\triangle L_{vi}(k)\\
&= \frac{1}{\gamma_{ci}} tr(\tilde{W}_{ci}^T(k+1)\tilde{W}_{ci}(k+1) - \tilde{W}_{ci}^T(k)\tilde{W}_{ci}(k))\\
&= \frac{1}{\gamma_{ci}} tr\Bigg(\Bigg(\tilde{W}_{ci}(k) + \gamma_{ci}\frac{\triangle\psi_{ci}(z_i(k))e_{ci}^T(k)}{\triangle\psi_{ci}^T(z_i(k))\triangle\psi_{ci}(z_i(k)) + 1}\Bigg)^T\Bigg(\tilde{W}_{ci}(k) + \gamma_{ci}\frac{\triangle\psi_{ci}(z_i(k))e_{ci}^T(k)}{\triangle\psi_{ci}^T(z_i(k))\triangle\psi_{ci}(z_i(k)) + 1}\Bigg)\\
&- \tilde{W}_{ci}^T(k)\tilde{W}_{ci}(k)\Bigg)\\
&= 2\tilde{W}_{ci}^T(k)\frac{\triangle\psi_{ci}(z_i(k))e_{ci}^T(k)}{\triangle\psi_{ci}(z_i^T(k))\triangle\psi_{ci}(z_i(k)) + 1} + \gamma_{ci}\frac{e_{ci}(k)\triangle\psi_{ci}^T(z_i(k))\triangle\psi_{ci}(z_i(k))e_{ci}^T(k)}{(\triangle\psi_{ci}^T(z_i(k))\triangle\psi_{ci}(z_i(k)) + 1)^2}\\
&= 2\tilde{W}_{ci}^T(k)\frac{\triangle\psi_{ci}(z_i(k))(-\tilde{W}_{ci}^T(k)\triangle\psi_{ci}(z_i(k)) - \triangle\varepsilon_{ci}(k))^T}{\triangle\psi_{ci}^T(z_i(k))\triangle\psi_{ci}(z_i(k)) + 1}\\
&+ \gamma_{ci}\frac{(-\tilde{W}_{ci}^T(k)\triangle\psi_{ci}(z_i(k)) - \triangle\varepsilon_{ci}(k))\triangle\psi_{ci}^T(z_i(k))\triangle\psi_{ci}(z_i(k))(-\tilde{W}_{ci}^T\triangle\psi_{ci}(z_i(k)) - \triangle\varepsilon_{ci}(k))^T}{(\triangle\psi_{ci}^T(z_i(k))\triangle\psi_{ci}(z_i(k)) + 1)^2}.
\end{aligned}
\tag{9.42}
$$

Substituting $e_{ci}(k)$ from (9.24) into (9.42) and using the C-S inequality, (9.42) is rewritten as

$$
\begin{aligned}
\triangle L_{vi}(k) \le &-\frac{2\tilde{W}_{ci}^T(k)\triangle\psi_{ci}(z_i(k))\triangle\psi_{ci}^T(z_i(k))\tilde{W}_{ci}(k)}{\triangle\psi_{ci}^T(z_i(k))\triangle\psi_{ci}(z_i(k)) + 1} - \frac{2\tilde{W}_{ci}^T(k)\triangle\psi_{ci}(z_i(k))\triangle\varepsilon_{ci}^T(k)}{\triangle\psi_{ci}^T(z_i(k))\triangle\psi_{ci}(z_i(k)) + 1}\\
&+ \frac{2\gamma_{ci}\tilde{W}_{ci}^T(k)\triangle\psi_{ci}(z_i(k))\triangle\psi_{ci}^T(z_i(k))\triangle\psi_{ci}(z_i(k))\triangle\psi_{ci}^T(z_i(k))\tilde{W}_{ci}(k)}{(\triangle\psi_{ci}^T(z_i(k))\triangle\psi_{ci}(z_i(k)) + 1)^2}\\
&+ \frac{2\gamma_{ci}\triangle\varepsilon_{ci}(k)\triangle\psi_{ci}^T(z_i(k))\triangle\psi_{ci}(z_i(k))\triangle\varepsilon_{ci}^T(k)}{(\triangle\psi_{ci}^T(z_i(k))\triangle\psi_{ci}(z_i(k)) + 1)^2}.
\end{aligned}
\tag{9.43}
$$

Based on Young's inequality $2a^T b \le ra^T a + \frac{1}{r} b^T b$ with $r > 0$, $\frac{\Delta\psi_{ci}^T(z_i(k))\Delta\psi_{ci}(z_i(k))}{\Delta\psi_{ci}^T(z_i(k))\Delta\psi_{ci}(z_i(k))+1} \le 1$ and $\frac{1}{\Delta\psi_{ci}^T(z_i(k))\Delta\psi_{ci}(z_i(k))+1} \le 1$, then the first difference of $L_{vi}(k)$ becomes

$$\begin{aligned}\Delta L_{vi}(k) \le & -(1-2\gamma_{ci})\frac{\tilde{W}_{ci}^T(k)\Delta\psi_{ci}(z_i(k))\Delta\psi_{ci}^T(z_i(k))\tilde{W}_{ci}(k)}{\Delta\psi_{ci}^T(z_i(k))\Delta\psi_{ci}(z_i(k))+1} \\ & + (1+2\gamma_{ci})\frac{\Delta\varepsilon_{ci}(k)\Delta\varepsilon_{ci}^T(k)}{\Delta\psi_{ci}^T(z_i(k))\Delta\psi_{ci}(z_i(k))+1} \\ \le & -(1-2\gamma_{ci})\frac{\Delta\psi_{c,m}^2}{\Delta\psi_{c,M}^2+1}\|\tilde{W}_{ci}(k)\|^2 + (1+2\gamma_{ci})\frac{1}{\Delta\psi_{c,M}^2+1}\Delta\varepsilon_{c,M}^2\end{aligned} \tag{9.44}$$

where $\Delta\psi_{c,m} \le \|\Delta\psi_{c,i}\| \le \Delta\psi_{c,M}$, which satisfies the PE condition.

Subsequently, the first difference of the last term in (9.33) is given by

$$\begin{aligned}\Delta L_{ui}(k) = & \frac{1}{\gamma_{ai}} tr(\tilde{W}_{ai}^T(k+1)\tilde{W}_{ai}(k+1) - \tilde{W}_{ai}^T(k)\tilde{W}_{ai}(k)) \\ = & \frac{1}{\gamma_{ai}} tr((\tilde{W}_{ai}(k) - \gamma_{ai}\psi_{ai}(\bar{z}_i)\hat{W}_{ci}^T(k)\psi_{ci}^{'}(\bar{z}_i)p_i[\hat{W}_{ci}^T(k)\psi_{ci}(\bar{z}_i)]^T)^T \\ & \cdot (\tilde{W}_{ai}(k) - \gamma_{ai}\psi_{ai}(\bar{z}_i)\hat{W}_{ci}^T(k)\psi_{ci}^{'}(\bar{z}_i)p_i[\hat{W}_{ci}^T(k)\psi_{ci}(\bar{z}_i)]^T) - \tilde{W}_{ai}^T(k)\tilde{W}_{ai}(k)) \\ = & -2\tilde{W}_{ai}^T(k)\psi_{ai}(\bar{z}_{i,s})\hat{W}_{ci}^T(k)\psi_{ci}^{'}(\bar{z}_i)p_i[\hat{W}_{ci}^T(k)\psi_{ci}(\bar{z}_i)]^T + \gamma_{ai}(\psi_{ai}(\bar{z}_{i,s})\hat{W}_{ci}^T(k)\psi_{ci}^{'}(\bar{z}_i) \\ & \cdot p_i[\hat{W}_{ci}^T(k)\psi_{ci}(\bar{z}_i)]^T)^T(\psi_{ai}(\bar{z}_{i,s})\hat{W}_{ci}^T(k)\psi_{ci}^{'}(\bar{z}_i)p_i[\hat{W}_{ci}^T(k)\psi_{ci}(\bar{z}_i)]^T).\end{aligned} \tag{9.45}$$

Due to $\hat{u}_i$ is admissible consensus control policy, $\hat{V}_i(k) = \hat{W}_{ci}^T(k)\psi_{ci}(\bar{z}_i)$ is finite, i.e., $\Gamma_m \le \|\hat{V}_i(k)\| \le \Gamma_M$. Then, the first difference of $L_{ui}(k)$ can be rewritten as

$$\begin{aligned}& \Delta L_{ui}(k) \\ & \le -2\Gamma_m \tilde{W}_{ai}^T(k)\psi_{ai}(\bar{z}_{i,s})\hat{W}_{ci}^T(k)\psi_{ci}^{'}(\bar{z}_i)p_i \\ & + \gamma_{ai}\Gamma_M^2(\psi_{ai}(\bar{z}_{i,s})\hat{W}_{ci}^T(k)\psi_{ci}^{'}(\bar{z}_i)p_i)^T(\psi_{ai}(\bar{z}_{i,s})\hat{W}_{ci}^T(k)\psi_{ci}^{'}(\bar{z}_i)p_i) \\ & \le -\Gamma_m\psi_{a,M}^2\|\tilde{W}_{ai}(k)\|^2 + \Gamma_m\nabla\psi_{c,M}^2\|\hat{W}_{ci}(k)\|^2 + \gamma_{ai}\Gamma_M^2\nabla\psi_{c,M}^2\psi_{a,M}^2\|\hat{W}_{ci}(k)\|^2 \\ & \le -\Gamma_m\psi_{a,M}^2\|\tilde{W}_{ai}(k)\|^2 + 2(\Gamma_m + \gamma_{ai}\Gamma_M^2\psi_{a,M}^2)\nabla\psi_{c,M}^2 W_{c,M}^2 \\ & + 2(\Gamma_m + \gamma_{ai}\Gamma_M^2\psi_{a,M}^2)\nabla\psi_{c,M}^2\|\tilde{W}_{c,i}\|^2.\end{aligned} \tag{9.46}$$

Combing (9.41), (9.44), and(9.45), the first difference of the Lyapunov function (9.33) at the triggering instants is given as

$$
\begin{aligned}
\triangle L_i(k) =& \triangle L_{\delta i}(k) + \triangle L_{vi}(k) + \triangle L_{ui}(k) \\
\le& -(1-2K)\|\delta_i(k)\|^2 + \sum_{j\in\bar{N}_i} 2s_i l_{ij}^2 g_M^2 (\psi_{a,M}^2 \|\tilde{W}_{aj}\|^2 + \varepsilon_{a,M}^2) \\
&- (1-2\gamma_{ci})\frac{\triangle\psi_{c,m}^2}{\triangle\psi_{c,M}^2+1}\|\tilde{W}_{ci}(k)\|^2 + (1+2\gamma_{ci})\frac{1}{\triangle\psi_{c,M}^2+1}\triangle\varepsilon_{c,M}^2 \\
&- \Gamma_m \psi_{a,M}^2 \|\tilde{W}_{ai}(k)\|^2 + 2(\Gamma_m + \gamma_{ai}\Gamma_M^2\psi_{a,M}^2)\nabla\psi_{c,M}^2 W_{c,M}^2 \\
&+ 2(\Gamma_m + \gamma_{ai}\Gamma_M^2\psi_{a,M}^2)\nabla\psi_{c,M}^2 \|\tilde{W}_{c,i}\|^2 \\
\le& -(1-2K)\|\delta_i(k)\|^2 + \sum_{j\in\bar{N}_i} 2s_i l_{ij}^2 g_M^2 \psi_{a,M}^2 \|\tilde{W}_{aj}\|^2 - \Gamma_m\psi_{a,M}^2\|\tilde{W}_{ai}(k)\|^2 \\
&- (1-2\gamma_{ci})\frac{\triangle\psi_{c,m}^2}{\triangle\psi_{c,M}^2+1}\|\tilde{W}_{ci}(k)\|^2 + 2(\Gamma_m + \gamma_{ai}\Gamma_M^2\psi_{a,M}^2)\nabla\psi_{c,M}^2\|\tilde{W}_{c,i}\|^2 + \zeta_i
\end{aligned}
\tag{9.47}
$$

where $\zeta_i = \sum_{j\in\bar{N}_i} 2s_i l_{ij}^2 g_M^2 \varepsilon_{a,M}^2 + (1+2\gamma_{ci})\frac{1}{\triangle\psi_{c,M}^2+1}\triangle\varepsilon_{c,M}^2 + 2(\Gamma_m + \gamma_{ai}\Gamma_M^2\psi_{a,M}^2)$ $\nabla\psi_{c,M}^2 W_{c,M}^2$.

From (9.47) and selecting $\gamma_{ci} < 1/2$, $\triangle L_i(k) < 0$ as long as

$$
\|\tilde{W}_{ci}(k)\| \ge \sqrt{(\triangle\psi_{c,M}^2+1)\zeta_i/((1-2\gamma_{ci})\triangle\psi_{c,m}^2 - 2(\Gamma_m+\gamma_{ai}\Gamma_M^2\psi_{a,M}^2)(\triangle\psi_{c,M}^2+1)\nabla\psi_{c,M}^2)}
$$

or

$$
\|\tilde{W}_{ai}(k)\| \ge \sqrt{(\zeta_i + 2s_i\sum_{j\in N_i} l_{ij}^2 g_M^2 \psi_{a,M}^2\|\tilde{W}_{aj}\|^2)/(\Gamma_m\psi_{a,M}^2 - 2s_i l_{ii}^2 g_M^2 \psi_{a,M}^2)}
$$

or

$$
\|\delta_i(k)\| \ge \sqrt{(\zeta_i/(1-2K)}.
$$

This means that the neighbor consensus errors, the weight estimation errors for the critic-action networks are UB. ■

Combined Case 1 with Case 2, we can obtain that the neighbor consensus errors, the weight estimation errors for the critic-action networks are UB.

Remark 9.4 The minimum inter-event time of event-triggered mechanism refers to the minimal time interval between two consecutive sampling instants, i.e., $\tau_i = k_{s+1}^i - k_s^i$, $s \in N$. In the discrete-time MASs, the minimal inter-event time is no less than the sampling time T_s, that is, $\tau_i \ge T_s$ so that the EPADP method for discrete-time MASs can avoid Zeno behavior.

9.4 Simulation Example

In this section, we give a numerical example to show the effective of ETADP method. Here, the communication graph consisting of four agents is considered as in Fig. 9.2. And the dynamics of each agent is represented as

Agent 1:

$$\begin{aligned}x_{11}(k+1) &= x_{12}(k) - x_{11}^2(k)x_{12}(k)\\ x_{12}(k+1) &= -(x_{11}(k) + x_{12}(k))(1 - x_{11}(k))^2 + x_{12}^2 u_1\end{aligned}$$

Agent 2:

$$\begin{aligned}x_{21}(k+1) &= x_{22}(k) - x_{21}^2(k)x_{22}(k)\\ x_{22}(k+1) &= -(x_{21}(k) + x_{22}(k))(1 - x_{21}(k))^2 + 1.5x_{22}^2 u_2\end{aligned}$$

Agent 3:

$$\begin{aligned}x_{31}(k+1) &= x_{32}(k) - x_{31}^2(k)x_{32}(k)\\ x_{32}(k+1) &= -(x_{31}(k) + x_{32}(k))(1 - x_{31}(k))^2 - 0.2x_{32}^2 u_3\end{aligned}$$

Agent 4:

$$\begin{aligned}x_{41}(k+1) &= x_{42}(k) - x_{41}^2(k)x_{42}(k)\\ x_{42}(k+1) &= -(x_{41}(k) + x_{42}(k))(1 - x_{41}(k))^2 + 0.3x_{42}^2 u_4.\end{aligned}$$

For the simulation purpose, we select the system parameters as follows. Let the matrices of the performance indices be $Q_i = 2I_2$, $R_i = 1$ and the gains of the critic-

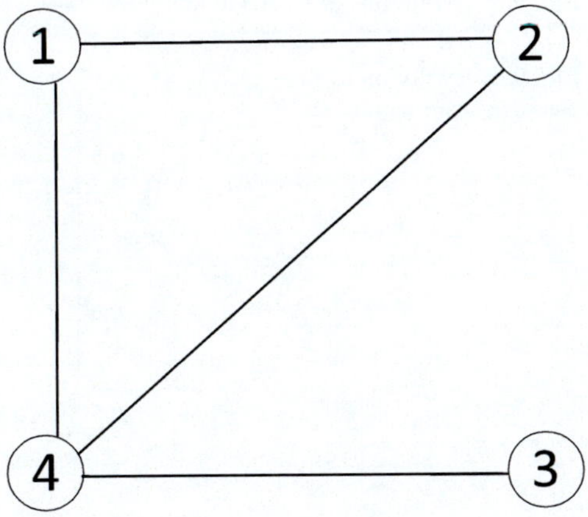

Fig. 9.2 Communication graph of four agents

action NNs are $\gamma_{ci} = 0.03$, $\gamma_{ai} = 0.05$. The critic-action weights are initialized randomly within [0, 0.01]. According to the adjacent relation in communication graph, we select the critic NN activation functions as:

$$\begin{aligned}
\psi_{c1} &= [\tanh(\delta_{11}^2) \quad \tanh(\delta_{12}^2) \quad \tanh(\hat{u}_1^2) \quad \tanh(\hat{u}_2^2) \quad \tanh(\hat{u}_4^2)]^T, \\
\psi_{c2} &= [\tanh(\delta_{21}^2) \quad \tanh(\delta_{22}^2) \quad \tanh(\hat{u}_1^2) \quad \tanh(\hat{u}_2^2) \quad \tanh(\hat{u}_4^2)]^T, \\
\psi_{c3} &= [\tanh(\delta_{31}^2) \quad \tanh(\delta_{32}^2) \quad \tanh(u_3^2) \quad \tanh(u_4^2)]^T, \\
\psi_{c4} &= [\tanh(\delta_{41}^2) \quad \tanh(\delta_{42}^2) \quad tanh(\hat{u}_1^2) \quad \tanh(\hat{u}_2^2) \quad \tanh(\hat{u}_4^2) \quad \tanh(\hat{u}_4^2)]^T.
\end{aligned}$$

And the action NN activation functions are

$$\psi_i = [\delta_{i1}^2 \quad \delta_{i1}\delta_{i2} \quad \delta_{i2}^2]^T, \quad i = 1, \ldots, 4.$$

The trajectories curves of the neighbor consensus errors for MAS are shown in Fig. 9.3 which indicates that MAS (9.1) can reach consensus. Meanwhile, Fig. 9.4 describes the 3-D phase plane plot of the states where all agents converge to the equilibrium point. In Fig. 9.5, optimal consensus control trajectories for four agents

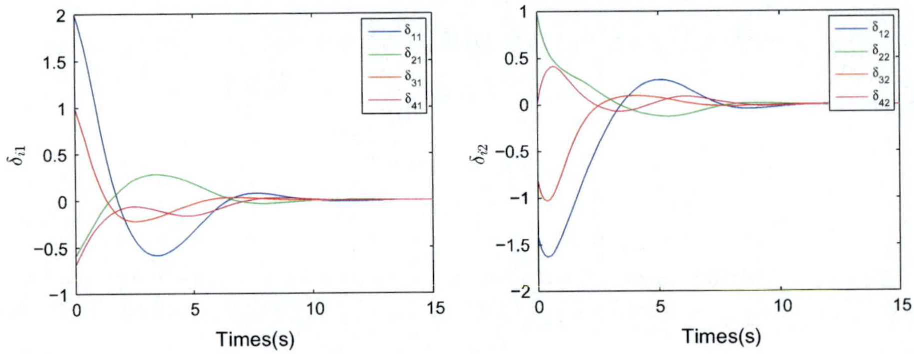

Fig. 9.3 Evolutions of the local neighbor consensus errors of four agents δ_i

Fig. 9.4 Evolution of the states for four agents

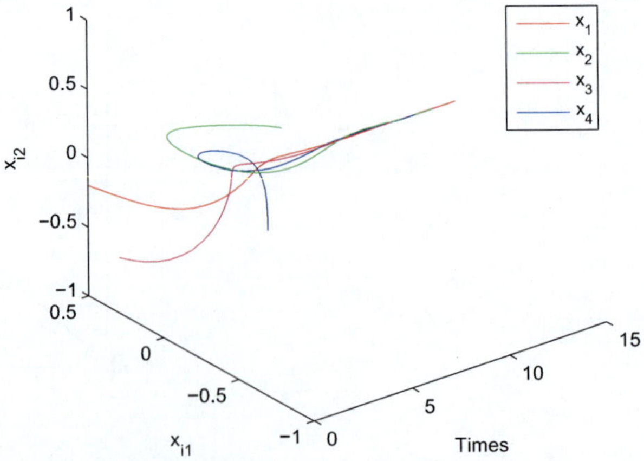

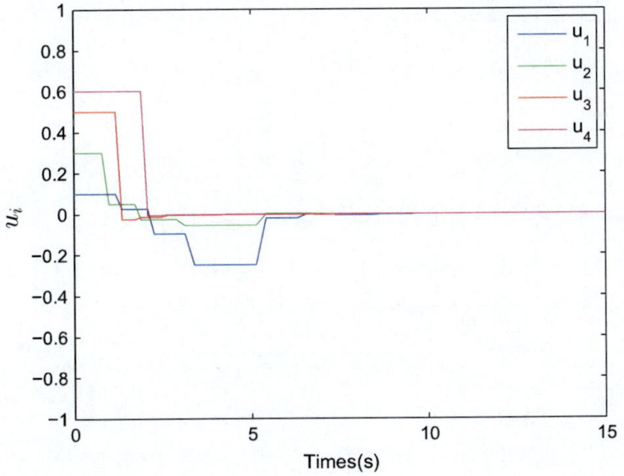

Fig. 9.5 Optimal consensus control for four agents

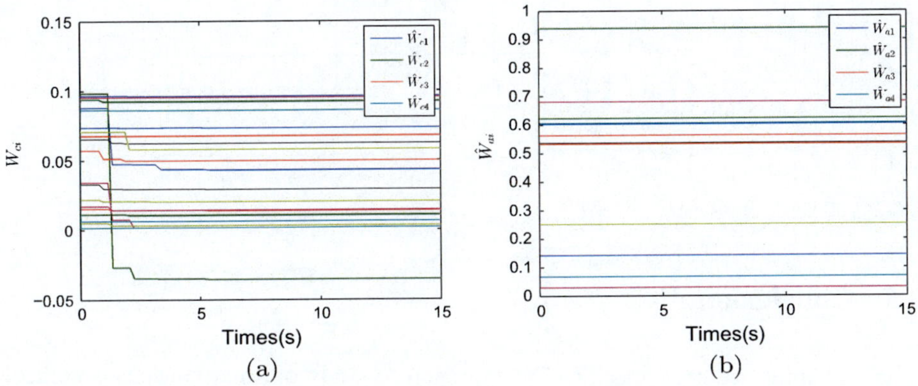

Fig. 9.6 Convergence of the weight estimations: **a** Critic NN; **b** Action NN

are shown. It is indicated from Fig. 9.5 that the control constraint $\|u_i\| \leq 1$ is satisfied under the ETADP method. Fig. 9.6 shows the convergence curves of the critic-action NN weight estimations, which implies that the weight estimation errors are UB. The event-triggered instants and inter-event time interval are shown in Fig. 9.7. It is observed that the number of optimal consensus control actions is largely reduced in the computation of around (19.5%) for the ETADP method than that of time-triggered ADP method. Thus, the proposed ETADP method can largely reduce the frequency of communication and decreases commutation cost.

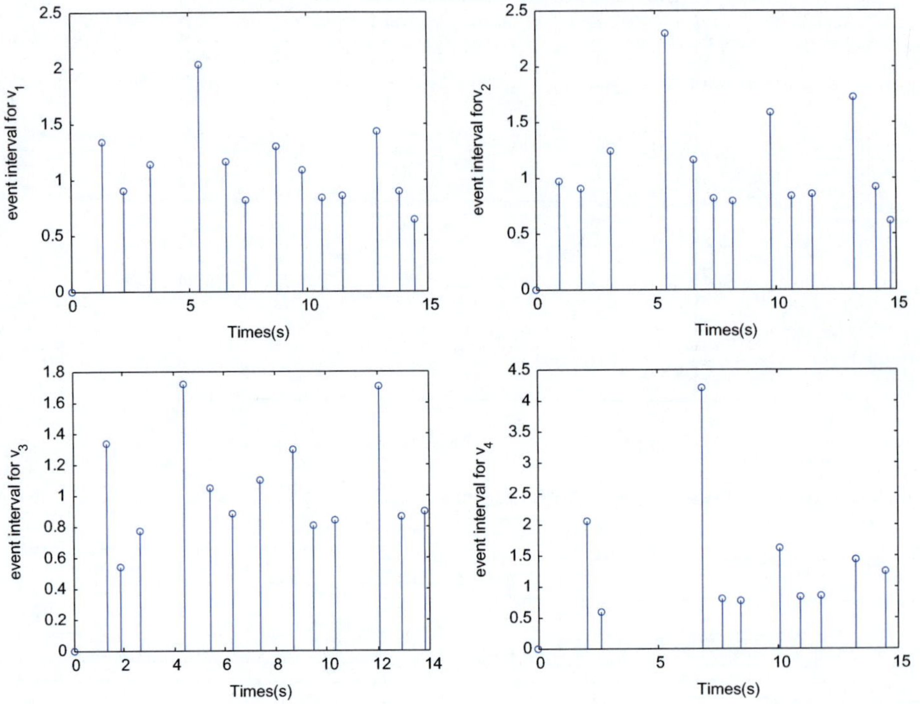

Fig. 9.7 Inter-event times of all agents

9.5 Conclusion

In this chapter, we used the ETADP approach to study optimal consensus control problem for nonlinear MASs with control constraints. The nonquadratic functions were introduced into the value functions to deal with control constraints and then the event-triggered constrained consensus control policies were derived by the HJB equations. The critic-action networks were designed to approximate the value functions and control policies by using the measurable system data, respectively. In the event-triggered sampling framework, the weights of the critic-action networks were only updated at the triggering instants. Furthermore, we proved the system stability and NN convergence using Lyapunov method. An example was given to show the effectiveness of the proposed ETADP method.

References

1. K.G. Vamvoudakis, F.L. Lewis, G.R. Hudas, Multi-agent differential graphical games: Online adaptive learning solution for synchronization with optimality. Automatica **48**(8), 1598–1611 (2012)
2. H. Zhang, J. Zhang, G.-H. Yang, Y. Luo, Leader-based optimal coordination control for the consensus problem of multiagent differential games via fuzzy adaptive dynamic programming. IEEE Trans. Fuzzy Syst. **23**(1), 152–163 (2015)
3. W. Zhao, R. Li, H. Zhang, Leader-follower optimal coordination tracking control for multi-agent systems with unknown internal states. Neurocomputing **249**, 171–181 (2017)
4. H. Zhang, D. Yue, W. Zhao, S. Hu, C. Dou, Distributed optimal consensus control for multiagent systems with input delay. IEEE Trans. Cybern. **48**(6), 1747–1759 (2017)
5. D. Liu, D. Wang, X. Yang, An iterative adaptive dynamic programming algorithm for optimal control of unknown discrete-time nonlinear systems with constrained inputs. Inf. Sci. **220**, 331–342 (2013)
6. H. Zhang, Y. Luo, D. Liu, Neural-network-based near-optimal control for a class of discrete-time affine nonlinear systems with control constraints. IEEE Trans. Neural Netw. **20**(9), 1490–1503 (2009)
7. X. Zhong, H. He, An event-triggered adp control approach for continuous-time system with unknown internal states. IEEE Trans. Cybern. **47**(3), 683–694 (2017)
8. A. Sahoo, S. Jagannathan, Stochastic optimal regulation of nonlinear networked control systems by using event-driven adaptive dynamic programming. IEEE Trans. Cybern. **47**(2), 425–438 (2017)
9. M.I. Abouheaf, F.L. Lewis, K.G. Vamvoudakis, S. Haesaert, R. Babuska, Multi-agent discrete-time graphical games and reinforcement learning solutions. Automatica **50**(12), 3038–3053 (2014)
10. A. Sahoo, H. Xu, S. Jagannathan, Near optimal event-triggered control of nonlinear discrete-time systems using neurodynamic programming. IEEE Trans. Neural Netw. Learn. Syst. **27**(9), 1801–1815 (2016)
11. J. Sarangapani, *Neural Network Control of Nonlinear Discrete-Time Systems* (CRC Press, 2006)
12. T. Dierks, S. Jagannathan, Online optimal control of affine nonlinear discrete-time systems with unknown internal dynamics by using time-based policy update. IEEE Trans. Neural Netw. Learn. Syst. **23**(7), 1118–1129 (2012)

Appendix
Algebraic Graph Theorys

In MASs, the information is exchanged among the agents through a communication network. In general, communication network can be modeled as a directed/undrecited graph, in which the agents are regarded as the nodes and the information flow between the agents are considered as the edges. In what follows, some basic concepts of algebraic graph theory are presented.

Denote $\mathcal{G} = \{\mathcal{V}, \mathcal{E}\}$ as a directed graph composed of a set of agents $\mathcal{V} = \{v_1, \ldots, v_N\}$, and a set of directed edges $\mathcal{E} = \{e_{ij} = (v_i, v_j) : v_i, v_j \in \mathcal{V}\} \subseteq \mathcal{V} \times \mathcal{V}$. An edge denoted as $e_{ij} = (v_i, v_j)$ means that node v_j can receive the information of node v_i, but not necessarily vice versa. In this case, v_i is called the parent node and v_j is the child node. It is assumed that self-edges (v_i, v_i) are not allowed. The set of neighbors of node v_i is denoted as $N_i := \{v_j | e_{ji} = (v_j, v_i) \in E\}$, and we define that $\bar{N}_i = \{N_i, i\}$ includes agent i and its neighbours N_i. A directed path from node v_i to node v_j is a edge sequence of (v_i, v_{i_1}),...,(v_{i_k}, v_j) which starts from node v_i and ends with node v_j. And v_i and v_j are called directed connected if there is a directed path from node v_i to node v_j. If all pairs of agents in $\mathcal{G}$ are directed connected, then $\mathcal{G}$ is called as directed connected graph. A digraph is said to be strongly connected if there is a directed path from node v_i to node v_j for all distinct nodes $v_i, v_j \in \mathcal{V}$. If there exists a root v_i, such that there is a directed path from the root to any nodes in the digraph, then it is called as a spanning tree.

Compared with directed graph, the pairs of nodes in undirected graph are unordered, i.e., $e_{ij} \in \mathcal{E}$ is equivalent to $e_{ji} \in \mathcal{E}$. A undirected path from node v_i to node v_j is a edge sequence of (v_i, v_{i_1}),...,(v_{i_k}, v_j) which starts from node v_i and ends with node v_j. And v_i and v_j are called undirected connected if there is a undirected path from node v_i to node v_j. If all pairs of agents in $\mathcal{G}$ are undirected connected, then $\mathcal{G}$ is called as undirected connected graph.

The adjacency matrix of a digraph $\mathcal{G}$ is denoted as $\mathcal{A} = [a_{ij}] \in R^{N \times N}$, where $a_{(i,j)} > 0$ if $(v_j, v_i) \in \mathcal{E}$ and $a_{(i,j)} = 0$, otherwise. And it is assumed that $a_{(i,i)} = 0$ for $\forall i \in V$. In an undirected graph, the definition of the adjacency matrix is similar except that $a_{ij} = a_{ji} foralli \neq j$. The in-degree matrix of $\mathcal{G}$ is given by $D = diag\{d_i\}_{i=1}^N$

D. Yue et al., *Distributed Cooperative Control and Communication for Multi-agent Systems*, https://doi.org/10.1007/978-981-33-6718-0

in which the in-degree d_i of node v_i is defined as the number of its neighbour nodes, i.e., $d_i = \sum_{j \in N_i} a_{ij}$. The Laplacian matrix of $\mathcal{G}$ can be acquired by $L = D - \mathcal{A}$.

Index

A
Action networks, 9, 173
Active disturbance rejection control, 5, 67
Actuator failures, 127
Adaptive Dynamic Programming (ADP), 4
Admissible consensus control policy, 160
Angular velocity, 111, 115, 116, 120, 124, 127, 128, 133, 134

B
Bellman's optimality principle, 8, 151, 161, 173

C
Communication delay, 5, 7, 17, 18, 27, 29, 33, 35, 38–40
Consensus control, 2, 4–9
Containment control, 2, 3, 6, 8, 127, 144, 148
Cooperative control, 2–4, 6–8, 109, 112, 115–117
Cost-to-go function, 174
Critic networks, 9, 173, 179

D
Data-driven control, 2, 3

E
Event generator, 176, 177
Event-Triggered Adaptive Dynamic Programming (ETADP), 173
Event-Triggered Control (ETC), 4

F
Formation tracking, 3, 5, 7, 67–69, 72, 74, 75, 78–81

H
Hamilton-Jacobi-Bellman (HJB), 8, 151, 173
Heterogeneous MASs, 2–5, 7
Hyperbolic tangent function, 175

I
Iterative Learning Control (ILC), 3

L
Leader-following consensus, 2, 18, 29, 35
Linear Matrix Inequality (LMI), 29, 32, 33
Local neighbor consensus error, 9, 152, 153, 162, 167, 173, 183
Lyapunov-Krasovskii function, 8, 67, 70, 73

M
Model-based control, 2, 3
Modified Rodriguez Parameter (MRP), 127
Monotonic odd function, 175
Multi-agent graphical games, 4
Multi-Agent System (MAS), 1

O
Optimal coordination control, 4, 181
Output regulation, 2, 4

D. Yue et al., *Distributed Cooperative Control and Communication for Multi-agent Systems*, https://doi.org/10.1007/978-981-33-6718-0

P
Parametric uncertainties, 127
Policy iteration, 8, 151, 163, 169
Positive definite matrix, 33, 86, 87, 93, 128, 174
Predictive control, 5, 7, 17, 20, 22, 23, 33, 34, 38–40

R
Reinforcement learning, 4
Rigid body, 109, 111, 112, 115, 116, 119, 120, 122, 124, 125

S
Sliding-mode control, 129, 139
Sliding-mode surface, 6, 129, 133, 134, 136–139, 142, 143

T
Time-triggered control, 4

U
Ultimately Bounded (UB), 181
Unmanned Aerial Vehicles (UAVs), 67

V
Value iteration, 4, 176, 181